Smart Agriculture

Volume 20

The book series Smart Agriculture presents progress of smart agricultural technologies, which includes, but not limited to, specialty crop harvest robotics, UAV technologies for row crops, innovative IoT applications in plant factories, and big data for optimizing production process. It includes both theoretical study and practical applications, with emphasis on systematic studies. AI technologies in agricultural productions will be emphasized, consisting of innovative algorithms and new application domains. Additionally, new crops are emerging, such as hemp in U.S., and covered as well. This book series would cover regions worldwide, such as U.S., Canada, China, Japan, Korea, and Brazil.

The book series Smart Agriculture aims to provide an academic platform for interdisciplinary researchers to provide their state-of-the-art technologies related to smart agriculture. Researchers of different academic backgrounds are encouraged to contribute to the book, such as agriculture engineers, breeders, horticulturist, agronomist, and plant pathologists. The series would target a very broad audience – all having a professional related to agriculture production. It also could be used as textbooks for graduate students.

Dilnoz Muhamediyeva · Lola Safarova ·
Gulom Uzakov · Arthur Gibadullin
Editors

Engineering and Digital Technologies in Agriculture

Springer

Editors
Dilnoz Muhamediyeva
Tashkent Institute of Irrigation
and Agricultural Mechanization Engineers
National Research University
Tashkent, Uzbekistan

Gulom Uzakov
Karshi State Technical University
Karshi, Uzbekistan

Lola Safarova
Samarkand State University of Veterinary
Medicine, Livestock and Biotechnologies
Samarqand, Uzbekistan

Arthur Gibadullin
Tajik Technical University named after
academician M. S. Osimi
Dushanbe, Tajikistan

ISSN 2731-3476 ISSN 2731-3484 (electronic)
Smart Agriculture
ISBN 978-981-92-0480-9 ISBN 978-981-92-0481-6 (eBook)
https://doi.org/10.1007/978-981-92-0481-6

This Springer imprint is published by the registered company Springer Nature Singapore Pte Ltd.
The registered company address is: 152 Beach Road, #21-01/04 Gateway East, Singapore 189721, Singapore

Preface

This book is about *Engineering and Digital Technologies in Agriculture.*

This book was created with the aim of developing and disseminating international experience in the field of engineering and information technology in agriculture.

A distinctive feature of the book is the inclusion of chapters written by authors from the Kyrgyz Republic, the Republic of Uzbekistan, the Republic of Tajikistan, the Republic of Turkmenistan, the Republic of Azerbaijan, and Russia—regions where agriculture is of significant economic importance.

Each chapter was reviewed by at least three reviewers. As a result of the reviewers' work, 11 of the 34 chapters were accepted for publication. The reviews were based on an assessment of the submitted articles' topic, the relevance of the research, the scientific significance and novelty, the quality of the materials, and the originality of the work. The authors were able to revise their papers and resubmit them for review.

The editors of *Engineering and Digital Technologies in Agriculture* express their gratitude to the staff of Springer, who supported the publication of this work. We would also like to thank the chapter authors, the reviewers, and everyone who helped organize this book and prepare this volume for publication in the Smart Agriculture series.

Tashkent, Uzbekistan
Samarqand, Uzbekistan
Karshi, Uzbekistan
Dushanbe, Tajikistan

Dilnoz Muhamediyeva
Lola Safarova
Gulom Uzakov
Arthur Gibadullin

Contents

Development of Innovative Technologies in Agriculture to Achieve Sustainable Development Goals 1
Firdavs Arabov, Farhod Sharofzoda, Kurbomad Sodikov, and Ramazoni Asomuddini

Neural Modeling in Agricultural Systems 9
Nazgul Seitkazieva, Talgarbek Niiazbekov, Asel Kartanova, Suiunbek Esenbai uulu, and Samara Zhusupbekova

Modeling Agricultural Problems Based on Neural Technologies 25
Ainura Dyikanova, Aida Turdubaeva, Turgunbek Zhumaliev, Sovetbek Mamytkanov, and Baratbek Sabitov

Reverse Engineering of Agricultural Machinery Components Using Combined 3D Scanning and Parametric Modeling Technologies 41
Maria Karelina, Maxim Pletnev, and Dmitriy Taldykin

Modeling Climate Impact on Agricultural Systems Using Machine Learning Methods 55
Sharshenbek Zhusupkeldiev, Chyngyz Sabitov, Rahat Sagyndykova, Damira Asanbekova, and Gulzat Ismailova

Application of Artificial Intelligence Methods in Precision Agriculture for Arid Regions 71
Narly Babanazarov, Ilmyrat Ilyasov, Serdar Dowletov, Guwanch Akyyev, and Gurbanbibi Orazbayeva

Investigation of the Interrelation of Physical and Mechanical Parameters of the Typical Surfaces Soil of Movement of Agricultural Tractors and Machines 87
Sergey Zhukov, Vladimir Makarov, and Alexander Belyaev

Fuzzy Mamdani Model of Diagnostics of Infectious Diseases in Cows 107
Dilnoz Muhamedieva, Lola Safarova, Khudaynazar Yunusov, and Rustam Kudratov

Crop Yield Forecasting and Water Stress Assessment Using Sentinel-2 and Machine Learning 119
Ilnur Miftakhov, Marat Ishbulatov, and Liana Miftakhova

Prospects for the Application of Digital Technologies to Ensure Food Security 135
Sevinch Ibadova, Rena Mamedova, Narmina Aliyeva, Durdana Shirinova, and Kamala Abdullayeva

Formal Description of Control Logic for the Palletizing Process in Agriculture 147
Mark Klyagin and Vladimir Kholopov

Development of Innovative Technologies in Agriculture to Achieve Sustainable Development Goals

Firdavs Arabov, Farhod Sharofzoda, Kurbomad Sodikov, and Ramazoni Asomuddini

Abstract This article examines the development of innovative agricultural technologies in the context of the transition to an agro-ecological system. The article examines the transition to an agro-ecological system and the introduction of innovative technologies into the cultivation and processing of agricultural products. The article identifies the main challenges faced by the agricultural sector of the Republic of Tajikistan over the past decade, including declining agricultural productivity, land erosion and degradation, water shortages for irrigation, rising groundwater levels, and soil salinization. Based on the author's research, the problems and lack of innovative infrastructure in the agricultural sector of the Khatlon region of the Republic of Tajikistan were identified. In this context, the author proposes ways to overcome existing problems in the process of improving the innovative infrastructure of agriculture.

Keywords Innovative technologies · Agriculture · Infrastructure · Sustainable development · Agro-ecological system

1 Introduction

In the context of global warming and climate change, it is impossible to imagine the ecological and economic development of agriculture without innovative technologies and the digitalization of its processes. This is because it is possible to overcome the challenges of sustainable agricultural development by implementing innovative technologies in the processing and use of natural resources, as well as protecting

F. Arabov (✉) · F. Sharofzoda
Research Institute for Sustainable Development and Green Economy, International University of Tourism and Entrepreneurship of Tajikistan, Dushanbe, Republic of Tajikistan
e-mail: firdavs300@mail.ru

K. Sodikov
Tajik National University, Dushanbe, Republic of Tajikistan

R. Asomuddini
National Academy of Sciences of Tajikistan, Dushanbe, Republic of Tajikistan

D. Muhamediyeva et al. (eds.), *Engineering and Digital Technologies in Agriculture*, Smart Agriculture 20, https://doi.org/10.1007/978-981-92-0481-6_1

aquatic ecosystems. The agricultural sector and its environmentally friendly production processes have acquired particular importance in the current economic environment due to the limited area of agricultural land, declining river flows, and acute water shortages during the hot season.

Renowned scientists Odinaev and Jonaloev [1] note that the trend of shrinking arable land in the Republic of Tajikistan and the Central Asian region, coupled with rising food prices, increasingly necessitates investment in the development of innovative infrastructure for sustainable agricultural development. Researcher Sheraliev and Teshaev [2] emphasize that sustainable development of the agricultural sector should be ensured through the integrated coordination of livestock and crop production, taking into account innovative infrastructure. International scientists Oriakhi and Arodoye [3] note that agriculture and its innovative development are essential for increasing household incomes and developing the country's socioeconomic sphere. It should be noted that agriculture and its infrastructure have played a key and decisive role in improving the living standards of the rural population and ensuring healthy nutrition in all countries. In this regard, in today's environment, innovative agricultural infrastructure is crucial for improving agricultural production, increasing crop yields, ensuring the profitability and viability of agricultural enterprises, and protecting aquatic ecosystems. Therefore, the development of a mechanism for state support for investment in innovative agricultural development and the creation of innovative infrastructure to improve agricultural landscapes, the natural ecosystem, the melioration of agricultural land, and the ecological status of natural resources, including land and water, has acquired particular significance in the Republic of Tajikistan.

2 Sources and Methodology

To assess the innovative infrastructure of the agricultural sector and the transition to sustainable development, as well as the effectiveness of its management, the author developed an indicator-based assessment algorithm. The information collected by the author allowed for an accurate assessment of the state of the innovative infrastructure in agriculture. Based on the conducted research, the necessary data was collected from agricultural enterprises through questionnaires to support the transition to an innovative system and the digitalization of the agricultural sector, which served as the basis for the accuracy of the research theory and methodology.

The effectiveness of the method used can be seen in the tables and figures presented. The discussion method allowed us to identify the process of withdrawal of agricultural land from agricultural use, the factors causing its erosion and degradation, as well as factors deteriorating the environmental situation in the agricultural sector.

3 Preliminary Results

To achieve the Sustainable Development Goals in agriculture, it is necessary to ensure the environmental sustainability of agriculture and the transition of the agricultural sector to an agro-ecological system [4]. To ensure the environmental sustainability of agricultural production, the agricultural infrastructure must transition to a new form of operation, namely, innovation and digitalization of the production process. In connection with the formation of an innovative infrastructure and the creation of conditions for doing business in the agricultural sector, an agrarian reform was carried out in the Republic of Tajikistan in 1997. Against this backdrop, a new form of management emerged, and the irrigation infrastructure of the agricultural sector was privatized, having fallen into disrepair and become unusable due to uncontrollable factors. Due to the deterioration of agricultural infrastructure, the ecological condition of irrigated agricultural lands deteriorated, leading to land salinization on an area of 210,000 ha [5]. Against the backdrop of low productivity of agricultural land and the unsatisfactory financial situation of agricultural enterprises, the process of introducing innovative technologies and strengthening the material and technical base of irrigated agriculture is not being updated.

Insufficient state budget funds and agricultural enterprises' insufficient revenues for land reclamation have led to a significant deterioration in the technical condition of drainage systems over the past decade. Analysis shows that in the republic's agricultural regions, the technical condition of 30% of inter-farm and 22% of on-farm drainage systems has deteriorated and become unusable [6]. Consequently, salinization of 60,000 ha of irrigated land in the republic is observed due to rising groundwater levels [7]. According to the analysis, if this trend continues over the next five years, the area of saline and low-productivity land in the country's agricultural regions is likely to reach 310,000 ha. According to statistics, due to the lack of necessary irrigation infrastructure, 18,956 ha of irrigated agricultural land were withdrawn from agricultural use in 2023, and 18,283 ha became fallow and wetlands [6]. In 2023, a total of 37,239 ha of land were withdrawn from agricultural use due to a lack of reclamation measures, a shortage of funds, and the absence of modern irrigation technologies, further hindering access to food security.

Therefore, to prevent land alienation, it is necessary to implement important reclamation measures, including ensuring backup water supplies and preventing land salinization during the off-season agricultural period, which requires additional irrigation water and, in turn, increases reclamation costs. Given limited access to fertile irrigated land, improving the reclamation condition of rocky soils is particularly important. These soils require additional crop cultivation, and their high permeability leads to significant losses of irrigation water.

At the same time, agriculture must transition to new methods of growing and processing agricultural products using innovative technologies and digitalization of the production process. According to the author's research in the Khatlon region of the Republic of Tajikistan, the agricultural infrastructure was found to be insufficiently prepared for the transition to digitalization. In 2023, only 2% of agricultural

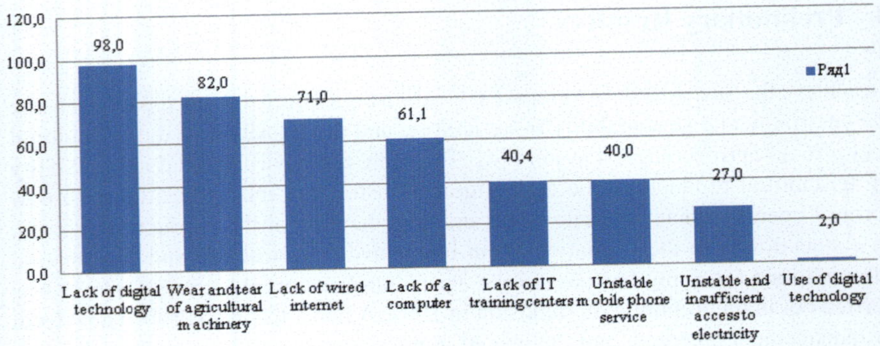

Fig. 1 Innovative infrastructure in agriculture in the Khatlon region

enterprises in Khatlon region used innovative and digital technologies, while 98% of enterprises did not employ innovative technologies (see Fig. 1).

Unstable and limited access to a constant power supply is observed in 27% of agricultural enterprises, and 40% lack high-quality mobile communications. 61.1% of surveyed enterprises lack computers and other digital technologies for administrative work. Many agricultural enterprises suffer from outdated agricultural machinery and technology, as well as a lack of new technologies. Existing equipment has high fuel costs, creates production costs and additional burdens for farmers, leading to increased production costs and reduced profitability.

According to the author's research and calculations, the costs of restoring agricultural infrastructure, such as rehabilitating on-farm drainage systems, cleaning drainage systems, introducing digital technologies into agricultural irrigation, and restoring irrigation machines, average 50,000–70,000 somoni (5000–7000 US dollars) per hectare of land in the Republic of Tajikistan, while land reclamation measures on one hectare of land require 20,000–30,000 somoni (2000–3000 US dollars).

Along with innovative factors, environmental, climatic, and anthropogenic factors also negatively impact agricultural production and crop yields in agricultural enterprises. As scientists Mukhabbatov and Kimsanov [8] emphasize, over the past decade, with the extensive development of the agro-industrial complex, the irrational use of natural resources and anthropogenic impact on the environment have led to degradation, pollution, swamping, the withdrawal of land from agricultural use, and the flooding of irrigated agricultural land. These factors can create obstacles to food security, job creation, and improved living standards.

Irrational use of water resources in agriculture also leads to rising groundwater levels, leading to salinization and reduced productivity of agricultural land. According to the UN World Water Assessment Programme report [9], which examines water use for sustainable development in agriculture, rural areas, and other economic sectors, water use and management across all sectors must be effectively implemented. In the Republic of Tajikistan, important steps are being taken to introduce innovative technologies in the cultivation and production of agricultural products with efficient use of water resources. As part of the implementation of innovative projects in the Kulyab region of the Republic of Tajikistan, a new irrigation method has been proposed to restore natural ecosystems and agricultural landscapes, as well as reduce the impact of climate change on desertification. According to research by Tagoeev [10] and other specialists, the Water-boxx technology was developed based on observations of natural processes and acts as a mechanism for retaining moisture from precipitation and snow, and is also ideal for growing plants and seedlings. According to experts [11], using Water-boxx allows seedlings to maintain moisture levels of 80–90% during the hot season. Furthermore, using 15 L of water can provide the necessary soil moisture for 30 days (approximately 30 ml per day). Tajikistan's agricultural sector is home to a large number of small agricultural enterprises and farms. Due to their poor financial situation, they are unable to attract investment for innovative projects. Therefore, the Water-boxx irrigation system, thanks to its low energy consumption (1–3 kW), can irrigate up to 3 hectares of orchards and vineyards. The cost of implementing the Water-boxx irrigation system is modest, amounting to $3000–5000. The system includes water pipes, solar panels, water pumps, and the Water-boxx system itself (Table 1).

According to the author's analysis, the economic efficiency of the Water-boxx system in potato cultivation is very high. If an innovative project costing $5000 were implemented on a potato cultivation area of 1.5 ha, the economic profit (land income) would be $2000 per year, meaning the project would have a payback period of 1.5 years.

$$1.5\ \text{ha} = 2000 \times 1.5 = 3000(\text{for potatoes}) \quad (1)$$

$$1.5\ \text{ha} = 6000 \times 1.5 = 9000(\text{for garden}) \quad (2)$$

Table 1 Application of Water-Box in growing seedlings in the Kulyab region

Region	Number of villages under control	Number of containers used, pcs	Land area, ha	Types of cultivated plants
Kulob	5	2100	3.80	Pistachio
Farkhor	5	1900	3.20	Pistachio
Hamadoni	4	1750	10.80	Pistachio
Vose'	15	6200	14.00	Pistachio
Danghara	26	9000	14.00	Pistachio
Total	55	20,850	34.80	

Source Annual report of the project "Development of livestock and pastures" in the Republic of Tajikistan for 2024

Table 2 Use of innovative technologies in the Kulyab region of the Republic of Tajikistan

Region	Name of the farmer	Land area, ha	Type of tree	Technical capacity		Total expenses, $	Yield per 1 ha, tons	Productivit, total tons	Total income in thousands of dollars $
				Solar panel, pc	Water pump, kWt				
Danghara	Farm "Abdusattor"	1.5	Apple Peach Abricot	14.0	2.0	5000.0	6.0 5.0 7.0	12.00 10.00 14.00	18.0 16.0 18.2
Bakjuvon	Farm "Mehrisniss"	2.0	Apple Pear Plum	12.0	2.4	6000.0	6.5 4.5 7.2	15.60 10.80 17.80	23.40 16.20 28.40

The profit in orchards is higher than in other agricultural crops, and the average annual profit from irrigating orchards with the Water-Box irrigation method is $6000, which on average brings the agricultural enterprise $9000 in profit from 1.5 ha of land, with a payback period of 1 year.

To adapt agriculture to climate change and ensure efficient use of water resources, an innovative irrigation method, Water-box, has been implemented on a 34.80-ha pistachio orchard in the Kulob district of the Republic of Tajikistan. As part of this project, 20,850 Water-boxx units have been installed on pistachio seedlings, providing soil moisture during the summer. This will increase crop yields and tree growth, ensure the development of natural ecosystems, and help agriculture adapt to climate change.

Water-boxx has been adopted by agricultural enterprises and farms in the Kulob district of the Republic of Tajikistan (Table 2).

According to the data in Table 2, the Abdusattor farm (DF) implemented an innovative seedling cultivation method. The seedlings were irrigated using a Waterboxx system, covering a total area of 1.5 ha. The technical implementation costs amounted to $5000. Within three years of receiving the seedlings, in the first year of harvesting, the farm's income amounted to $52,200, demonstrating the high cost-effectiveness of implementing the innovative irrigation methods.

4 Discussion

One of the main challenges in the transition to an agro-ecological system and the digitalization of the country's agriculture is the lack of advisory centers, insufficient methodological support, and insufficient investment. Demishkevich G. M. and Petrov A. A. believe [12] it is necessary and timely for agricultural producers and enterprises to study best practices, new irrigation methods, and the use of innovative technologies and digitalization of the production process, as well as the development of agriculture based on these practices in modern conditions. The use of innovative technologies, new irrigation methods, and crop cultivation improves the economic efficiency of enterprises, significantly increases agricultural yields, and facilitates the adaptation of the agricultural sector to climate change.

According to research conducted by scientists at agricultural enterprises, innovative crop cultivation methods have reduced administrative costs by up to 30%, water consumption by up to 65%, and increased resource efficiency and crop yields by up to 18% [13]. In this regard, to improve the production process and the environmental condition of agricultural lands in modern conditions, it is first of all necessary to develop innovative infrastructure in the agricultural sector and modern methods of cultivating agricultural crops.

5 Conclusion

To achieve the SDGs and the Government of the Republic of Tajikistan's goals for sustainable development in agriculture, food security, improving living standards, and exporting environmentally friendly products to regional and global markets, the agricultural sector must be provided with the necessary innovative infrastructure. In this context, in agricultural regions, increasing the efficiency of agricultural production through the introduction of digital and innovative technologies based on state support for the private sector through the "state-science-business" mechanism can take the agricultural sector to a new level of production. The creation of technology parks in agricultural regions will provide farmers with the opportunity to learn about new technological advances and irrigation methods, as well as work to improve agricultural infrastructure. This factor has a synergistic effect on the development of the socio-economic and production spheres of agriculture, ensuring the innovative development of the agricultural sector in subsequent stages. According to scientists Marina and Elena [14] the creation of agro-technical parks will contribute to the improvement and modernization of the infrastructure of the regional agricultural sector, simultaneously increasing the efficiency of agricultural production and having a positive impact on the economic development of the region. At the same time, innovative agricultural infrastructure will help reduce anthropogenic impact on the environment, preserve agricultural landscapes, and improve the ecological state of aquatic and natural ecosystems.

References

1. Odinaev, Kh.A., Jonaloev, M.S.: Development of Ecological and Economic Regulation of Land Use in an Arid Region, p. 175. Tajik National University, Dushanbe (2023)
2. Sheraliev, E.N., Teshaev, M.S.: Theoretical aspects of the formation and development of the mechanism of sustainable land use in arid regions. Finan. Econ. Bull. **4**(28), 121–126 (2021)
3. Oriakhi, D.E., Arodoye, L.N.: The government size–economic growth relationship: Nigerian econometric evidence using a vector autoregression model. Int. J. Bus. Manage. **8**(10), 126 (2013)

4. Sodikov, K.A., Arabov, F.P.: Digitalization as a key factor of sustainable development of agriculture in Tajikistan. In: Integration of Sciences—2021: Proceedings of the International Scientific and Practical Conference, Krasnodar, July 20, 2021. – Krasnodar: Russian Energy Agency of the Ministry of Energy of Russia, Krasnodar CNTI Branch, pp. 85–90 (2021)
5. State Program for the Development of New Irrigated Lands and Restoration of Lands Withdrawn from Agricultural Use in the Republic of Tajikistan for 2022–2027, Dushanbe, p. 90 (2022)
6. Agency for Land Reclamation and Irrigation under the Government of the Republic of Tajikistan. Available at: https://www.alri.tj/. Accessed 19 April 2025
7. Zieev, B.S.: Development of institutional foundations of ecological and economic regulation of land use, Bulletin of the Tajik National University. Ser. Social Econ. Social Sci. **7**, 74–80 (2023)
8. Muhabbatov, Kh.M., Kimsanov, U.O.: Formation and development of the green economy in Central Asian countries: ecological, economic and social aspects, Bulletin of Bokhtar State University named after Nosir Khusrav. Ser. Human. Econ. Sci. **1–1**(83), 217–221 (2021)
9. World Water Assessment Programme: The United Nations World Water Development Report 3: Water in a Changing World. UNESCO, Paris and Earthscan, London (2009)
10. Tagoev, J.H., Saidov, P.M.: Development of innovations in farms and enterprises of the agro-industrial region of Kulob. Finan. Econ. Bull. **2**(31), 56–64 (2022)
11. Vishnevsky, V., Dementyev, V.: Innovations, institutions and evolution. Probl. Econ. **9**, 41–62 (2010). https://doi.org/10.32609/0042-8736-2010-9-41-62
12. Demishkevich, G.M., Petrov, A.A.: Formation of the institution of agricultural consulting as a priority direction for the development of innovative infrastructure of the agro-industrial complex. Agri-Food Policy Russia **4**, 42–45 (2012)
13. Khokhlov, R.R., Degtyarev, B.N., Stelmak, N.E., Safonov, V.A., Bulgarov, M.A.: Modern methods for the development of production infrastructure of the agro-industrial complex of the Krasnodar Territory. Innov. Econ. Inform. Anal. Forecasts **6**, 93–99 (2024). https://doi.org/10.47576/2949-1894.2024.6.6.013
14. Nechaeva, M.L., Demidova, E.E.: Development of social and production infrastructure of agriculture. Bull. NGIEI **8**(159), 96–107 (2024). https://doi.org/10.24412/2227-9407-2024-8-96-107

Neural Modeling in Agricultural Systems

Nazgul Seitkazieva, Talgarbek Niiazbekov, Asel Kartanova, Suiunbek Esenbai uulu, and Samara Zhusupbekova

Abstract This paper examines the problem of modeling complex systems using neural modeling. A neural network based on transfer learning of big data was constructed for predicting viral and bacterial diseases in poultry. To optimize the neural network, it was proposed to tune the hyperparameters of the ResNet and VGG 16 models based on modifications to the learning rate and regularization. Disease recognition models for coccidiosis, salmonellosis, and Newcastle disease were constructed with high prediction accuracy, precision, and error. A comparative analysis of models built using transfer learning before and after hyperparameter tuning was conducted. Convolutional neural networks (CNNs) and deep learning models were created for predicting poultry diseases by classifying healthy and unhealthy feces for three disease types. Unhealthy fecal images, which may be symptoms of coccidiosis, salmonellosis, and Newcastle disease, were classified. Models were built using the basic CNN architecture of various configurations. Models were trained using images of bird feces labeled at a farm and during laboratory testing, and then fine-tuned using neural network hyperparameters. The Adam optimizer was used to create the CNN, Resnet, and Efficient Net models for bird disease recognition. Results were obtained for the Resnet model with the Adam optimizer, with and without hyperparameter tuning. Test accuracy results without fine-tuning were 85.06% for the base CNN and 87.85% for the updated neural network with hyperparameter tuning. Fine-tuning with freezing the transfer learning and normalization layers improved the model accuracy to 95.01%, with F1 scores for all classifiers above 83% across all four classes. Prediction reports, confusion matrices, and ROC-AUC curves were plotted for the Efficient Net and Resnet models.

N. Seitkazieva (✉) · T. Niiazbekov
Kyrgyz State University named after I. Arabaev, Bishkek, Kyrgyzstan

A. Kartanova
Kyrgyz National University named after J. Balasagyn, Bishkek, Kyrgyzstan

S. Esenbai uulu
Osh State University, Osh, Kyrgyzstan

S. Zhusupbekova
Kyrgyz National Agrarian University named after K.I. Scriabin, Bishkek, Kyrgyzstan

D. Muhamediyeva et al. (eds.), *Engineering and Digital Technologies in Agriculture*, Smart Agriculture 20, https://doi.org/10.1007/978-981-92-0481-6_2

Keywords Agricultural systems · Poultry farming · Poultry diseases · Disease symptoms · Deep learning · Convolutional neural networks · Modeling · Adama optimizer · Forecasting · Model accuracy · Fine-tuning

1 Introduction

Early detection of poultry diseases on poultry farms, due to increasing bird populations, is costly, requiring specialists and laboratory analysis, which in many cases is expensive. Climate change, fluctuating temperatures, and environmental conditions complicate poultry farming in rural and urban settings. Especially recently, abnormal weather conditions in summer and winter complicate the birds' environment. Viral diseases are rapidly spreading in poultry farming. Therefore, this study explores specific modeling approaches based on deep learning methods for an in-depth analysis of the model structure and its accuracy. Classic poultry diseases such as Salmonellosis, Newcastle, and Coccidiosis are experiencing new patterns of spread. Early detection is the primary preventative measure for detecting poultry diseases. The use of artificial intelligence tools, such as deep learning methods, in this area enables modeling the process for the early detection of poultry diseases. In this paper, a prediction problem is solved using deep learning technologies based on poultry feces, the purpose of which is to recognize feces images collected at different times using deep learning methods and computer vision technologies.

The constructed models take into account multiple parameters, including climatic and agricultural factors in poultry farming, for the purpose of predicting diseases, which can be detected early in poultry production with minimal costs. The advantages of using modern deep learning models, which are a modern approach to identifying the root causes of many viral and bacterial diseases, are demonstrated. Highly accurate models were obtained using convolutional neural networks and transfer learning models such as ResNet, VGG16, and MobileNet. B EfficientB4. The effectiveness of this technology in poultry farming can significantly increase poultry meat and egg production [1]. Creating models with high accuracy, early detection, and high-probability forecasting will undoubtedly increase gross output and household income, and its growth in the poultry industry is discussed in [2, 3].

Currently, due to population growth and demand, it is necessary to adjust and increase the quantity and quality of poultry [4, 5]. To prevent mass poultry mortality and significant economic losses, a thorough preliminary analysis of these diseases and modern methods for early detection are essential for preserving poultry flocks. Therefore, accurate predictive models are needed that can be used to manage farms to increase or decrease poultry production [6–8].

Laboratory studies [9, 10] and poultry farming practices show that the most common chicken diseases include avian cholera, helminth infections, salmonellosis, avian coccidiosis, and Newcastle disease. This requires a special approach to constructing detection models for these diseases, taking into account the signs of their occurrence. Practical studies [11, 12] based on artificial intelligence methods have shown that salmonellosis and coccidiosis, as gastrointestinal infections, and infected birds can be modeled with 100% accuracy using neural network training.

Newcastle disease [13] is a significant, widespread disease of poultry. It is considered one of the most serious diseases of poultry, affecting the respiratory, neurological, and digestive systems of poultry. Modeling this disease requires additional data requirements. Multimodal data is required for model training, which makes data collection challenging. This disease is very insidious and fatal once infected, with many birds and poultry dying without showing any symptoms. Various methods have been used to predict this disease [13, 14].

In this paper, the problem is solved using various transfer learning neural network architectures and a method for recognizing poultry diseases using a neural network architecture, including a modern deep learning image recognition method, and other advanced deep learning approaches such as CNN, Mobile Net V2, VGG16, and Efficient. The Adam and RMSProp optimizers serve as neural network optimizers in constructing all models. However, in some cases, testing deep learning models for disease diagnosis requires datasets annotated with laboratory methods such as polymerase chain reaction (PCR).

2 Materials and Methods

The basic mathematical apparatus of neural modeling relies primarily on the mathematical apparatus of convolution. Convolutional neural networks (CNNs) are a specialized class of neural networks that effectively work with images, audio, and other data with spatial or temporal structure. Their key feature is the use of convolutional layers, which automatically detect local patterns. Basic mathematical operations in CNNs. Convolution is an operation applied to input data (images, signals) using a filter (kernel). The formula for discrete convolution for a 2D image is defined as follows.

$$(S^{*}K)(i,j) = \sum_{m}\sum_{n} S(i-m, j-n) \cdot K(m,n), \quad (1)$$

where S is the input matrix (image), K is the convolution kernel (filter), and (i, j) are the coordinates of the output value. For example, a 3×3 filter is applied to an image, calculating a weighted sum of pixels in the neighborhood. Pooling is a dimensionality reduction operation that preserves important features. Max Pooling:

$$\mathrm{P(i,j)} = \max_{(x,y)\in R_{i,j}} S(x,y), \quad (2)$$

where R_{ij} is a local image region (e.g. 2×2). Average Pooling is defined as follows

$$\mathrm{P}(i,\mathrm{j}) = \frac{1}{|R_{i,j}|}\sum_{(x,y)\in R_{i,j}} S(x,y) \quad (3)$$

Activation functions are an essential element of neural networks for creating nonlinear models. They are essential for identifying nonlinear relationships in complex data. Here are examples of activation functions used in neural networks.

ReLU (Rectified Linear Unit):

$$f(x) = \max(0, x), \tag{4}$$

Sigmoid:

$$\sigma(x) = \frac{1}{1 + e^x} \tag{5}$$

Softmax (for classification):

$$\text{softmax}(x_i) = \frac{e^{x_i}}{\sum_{j=1}^{N} e^{x_j}}. \tag{6}$$

Backpropagation is part of training a convolutional neural network (CNN). CNNs are trained using gradient descent, minimizing a loss function. The loss function (Cross-Entropy for classification) is:

$$L = -\sum_i y_i \log(\hat{y}_i), \tag{7}$$

where y_{i} is the true class, and $\hat{y}_i$ are the predicted values. The gradient descent weighting is updated using recurrence relations.

$$W_{new} = W_{old} - \eta \cdot \frac{\partial L}{\partial W} \tag{8}$$

where η is the learning rate. L is the loss function (7).

The architecture of convolutional neural network (CNN) elements includes convolutional layers, which extract features, pooling layers, which reduce the dimensionality of matrices, and classification using fully connected layers. Dense and dropout regularization are powerful tools for neural networks, the main purpose of which is to prevent overfitting of models. Examples of convolutional neural networks (CNN) include LeNet-5, ResNet, VGG, and LeNet-5. In these models, convolutional network technology is defined as a sequence of operations: Convolution → Pooling → Convolution → Pooling → FC → Output. The ResNet model uses skip connections, which is defined as y = F (x) + x. The structure of CNNs is based on convolutions (local data processing).

Pooling (dimensionality reduction), nonlinear activations, and backpropagation learning. These methods allow for the efficient processing of images, videos, and other structured data. The main advantage of any deep learning method is its ability to utilize large data sets, which allows it to model accurate models. One important method for obtaining large data sets is data augmentation. Below are images of the main infectious diseases of poultry.

Data upscaling and downscaling can increase the accuracy of models, bias, and a combination of several data augmentation methods. Various neural network optimizers are widely used in the work to build poultry disease recognition models. The main purpose of optimizers in deep learning is to configure model parameters to minimize the loss functional. The choice of optimizer can greatly affect the performance of the model and the speed of training. In this work, we used the advanced Adam optimizer, which incorporates properties of many other algorithms. Adam (short for "adaptive momentum estimation") is an optimization algorithm that combines ideas from stochastic gradient descent with momentum and RMSProp. This algorithm is a fundamental algorithm and is used in many deep learning architectures. Similar to RMSProp, it uses an exponentially decaying mean of gradients and squared gradients to determine the updated scale. It also uses momentum to help the optimizer navigate the loss function more efficiently. The recurrence formulas in this optimizer are defined by the formulas as follows:

$$g = \nabla_{\theta} \mathrm{L}\left(\theta; x^{(i)}; y^{(i)}\right) \tag{9}$$

$$G = \beta G + (1 - \beta) \cdot g \odot g \tag{10}$$

$$\theta = \theta - \frac{\alpha}{\sqrt{G + \varepsilon}} \odot g \tag{11}$$

where g is the matrix of squared gradients, ε is a small constant added to avoid division by zero, and β is the decay rate hyperparameter. This algorithm works particularly well with sparse data and is the basis of the Adam optimizer. Let us consider the mathematical formulation of the minimization problem with the Adam optimizer. The mathematical formulation of the minimization problem with the Adam optimizer for the loss function $J(\theta)$, where $\theta \in R^d$ is the unknown parameter of the model $J(\theta)$ is the differentiable loss function for the minimization problem. The problem is to find the parameter θ^* that should minimize the following quality functional $J(\theta)$:

$$\theta^* = argmin_{\theta}\, J(\theta) \tag{12}$$

Let us now describe the process of constructing the Adam algorithm for this problem. The optimizer combines the ideas of adapting the learning rate RMSProp and accelerating the learning rate, momentum. The parameters included in the optimizer are the following: α is the initial learning rate of the neural network, β_1 and $\beta_2 \in [0, 1]$, are included as moving average coefficients; in the calculation formulas, $\beta_1 = 0.9$ and $\beta_2 = 0.999$ are usually taken. For numerical stability of the recurrent neural network, a small constant $\varepsilon = 10^{-8}$ is taken. The process of constructing a recurrent minimizing sequence is defined as follows.

Initialization of the model parameters: the first moment $m_0 = 0$, which estimates the gradient. The second moment $\upsilon_0 = 0$, is included as the estimate of the squared gradient. t = 0 is the iteration counter. To update the optimizer parameters at each iteration t, we perform the following steps.

Updating gradients:

$$g_t = \nabla_\theta L(\theta_t), \tag{13}$$

Using the following recursive formula, we update the average gradients—the first moment:

$$m_t = \beta_1 m_{t-1} + (1 - \beta_1)\, g_t \tag{14}$$

The recurrence formula for updating the second moment is determined by the formula:

$$\upsilon_t = \beta_2 \upsilon_{t-1} + (1 - \beta_2)\, {g_t}^2 \tag{15}$$

Correction of moments is very important:

$$\hat{m}_t = \frac{m_t}{1 - {\beta_1}^t}, \hat{\upsilon}_t = \frac{\upsilon_t}{1 - {\beta_2}^t}, \tag{16}$$

The sequence minimizing function taking into account the Adam optimizer is determined by the relations (8):

$$\theta_{t+1} = \theta_t - \alpha \cdot \frac{\hat{m}_t}{\sqrt{\hat{\upsilon}_t} + \varepsilon} \tag{17}$$

This method calculates the adaptive learning rate for each parameter. The Adam optimizer stores both the decaying mean of past gradients, similar to momentum, and the decaying mean of past squared gradients.

Thus, it combines the advantages of both methods.

This algorithm converges faster than other optimization algorithms and performs well with uncertain data. The momentum, m_t, is analogous to momentum and reduces fluctuations in the direction of an increasing gradient. The second moment, υ_t, adaptively scales the model's learning rate for each parameter. It corrects biases. When $\beta_1 = 0$, Adam transforms into RMSProp. When $\beta_1 = 0$ and $\beta_2 = 0$, it transforms into regular gradient descent. When using the Adam optimizer, the model's learning rate adapts for each parameter and converges quickly, including noisy data. This optimizer also performs well with large and sparse data.

Transfer learning is a powerful tool for building nonlinear models. This research utilizes one of the modern approaches to model building—transfer learning—whose primary goal is to use pre-trained models on big data. Unfreezing the initial layers built by this technology and further training the remaining layers of the neural network allows for the construction of a high-quality transfer learning model. This process is particularly effective when image data is insufficient for training models on big data. Transfer learning models can be effective when using various types of neural network optimizers, as well as after tuning the network's hyperparameters. Transfer learning is a machine learning method in which knowledge gained from solving one problem is transferred to another, similar task. This is especially useful when data for the new task is scarce. In neural networks, transfer learning typically involves: Using a pre-trained model, such as VGG or ResNet. Fine-tuning a portion of the network on new data. As discussed, freezing some layers is necessary to preserve previously extracted features. The mathematical formulation of transfer learning relies on the following general problem statement: a source task with data $M_s = \{X_S, Y_S\}$, a model f_S, and a target task with data $M_T = \{X_T, Y_T\}$, where $|M_T| << |M_S|$.

The goal is to tune the f_T model using knowledge from f_S to minimize the error on M_T. Feature Extraction Transfer (FET)/base layers of the pretrained f_S model are used as a feature extractor. A new classifier g is trained on top of these features. Formally, we have the following model expression:

$$f_T\ (\mathrm{x}) = g\left(f_S^{feat}(x)\right) \tag{18}$$

where f_S^{feat} are frozen layers.

During fine-tuning, some of the f_S layers are unfrozen and retrained on the M_T set. In this case, we optimize the error functionals constructed based on the M_s and M_T sets using the following combination:

$$\mathcal{L} = \alpha\mathcal{L}_S + (1 - \alpha)\mathcal{L}_T \tag{19}$$

where α is the transfer coefficient. The next important element of transfer learning is regularization, which prevents previous training values from being reset. During retraining, the model may not take this old knowledge into account. In this case, L_2 regularization is used, which penalizes large changes in weights. Here is the penalty formula.

$$\mathcal{L}_{reg} = \lambda||\theta - \theta_S||_2^2 \tag{20}$$

Other regularization methods include L_1, which resets many weights, and elastic net, which aims to preserve important weights. The main advantage of transfer learning for solving practical problems is layer freezing, whereby the early layers of a CNN (edge and texture detection) remain unchanged, while the later layers (classifier) are retrained. The neural network's learning rate is varied, and the model gradually adds new blocks while preserving the old ones.

3 Results and Discussion

All models constructed below use the Adam optimizer. RMSProp was used to compare model quality. Specific practical examples of poultry disease recognition were examined. A multi-class classification problem was considered, identifying the following classes of diseases in poultry: Healthy, Salmonellosis, Newcastle, and Coccidiosis. CNN models were constructed with the Adam and RMSProp optimizers, with a learning rate of 0.001 and neural network architecture. To achieve maximum model performance and accuracy, the key hyperparameters of the neural network were optimized. Table 1 summarizes the key points for step-by-step improvement of model quality. A detailed table of key hyperparameters for neural network tuning is also provided.

Throughout the neural network architecture settings, we will use the data in Table 1. The VGG16 neural network hyperparameters for this class were tuned using the following key parameters. The learning rate ranges from 10^{-2} to 10^{-4}, and the batch size is 32. In case of overfitting, the model uses network regularization using the Dropout method of 0.5. The case of large batch sizes (> 32) was not considered. Full details of the VGG16 transfer learning model implementation for poultry disease recognition are presented below. Data augmentation was performed on the training set. Results of training the VGG 16 model with the tuned parameters (Figs. 1 and 2).

Table 1 Key hyperparameters for fine-tuning neural network architecture

Hyperparameter	Possible values	Description
Learning rate	0.001, 0.0001, 0.00001	Learning speed
Batch size	16, 32, 64	Batch size
Dropout rate	0.3, 0.5, 0.7	Regularization
Dense units	128, 256, 512	Number of neurons
Optimizer	Adam, SGD, RMSProp	Optimization algorithm

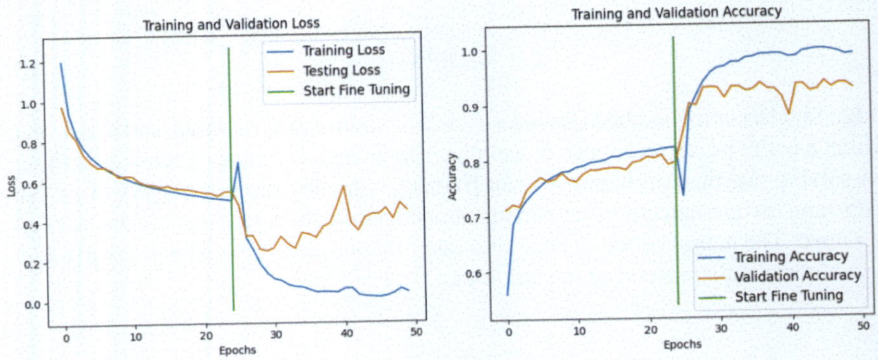

Fig. 1 VGG 16 transfer learning model after hyperparameter tuning on training and test data

Model : VGG16 Fine Tuning Model

	precision	recall	f1-score	support
cocci	0.98	0.93	0.95	247
healthy	0.87	0.95	0.91	276
ncd	0.91	0.91	0.91	54
salmo	0.95	0.91	0.93	239
accuracy			0.93	816
macro avg	0.93	0.92	0.92	816
weighted avg	0.93	0.93	0.93	816

Accuracy score: 0.93
F1 score average: 0.92

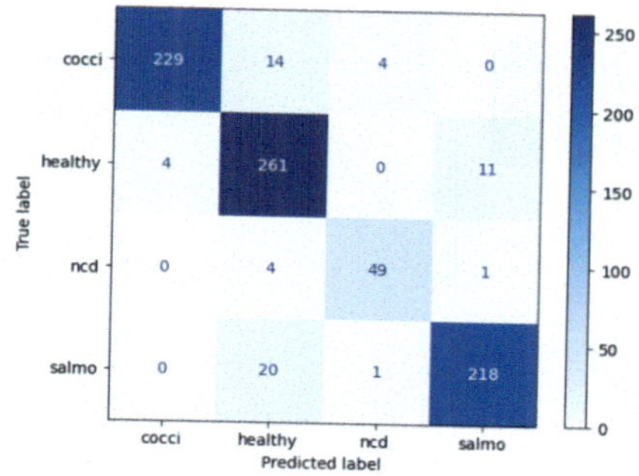

Fig. 2 Model prediction results and confusion matrix of VGG16 transfer learning with hyperparameter tuning

In this case, the VGG16 model achieved transfer learning performance with an accuracy of up to 98%. Figure 3 shows the MobileNetV2 model's transfer learning prediction results after hyperparameter tuning. The error matrix for the MobileNetV2 model for poultry diseases is calculated. Error and accuracy plots for the MobileNetV2 model after hyperparameter tuning are also plotted.

After hyperparameter tuning, the MobileNetV2 model's performance and validation on the test dataset achieved an accuracy of 0.93 and an F1 score of 0.90. Overall, the predicted values and comparison reports for the CNN, MobileNet V2, and VGG16 models before and after hyperparameter tuning show good results (Figs. 4, 5, and 6).

Using various neural modeling architectures is important when implementing neural networks and creating models. The EfficientB4 transfer learning model is an effective model for complex image data. We now present the results of implementing the EfficientB4 model with the Adam optimizer for predicting poultry diseases in the Newcastle Disease (Salmonellosis, Coccidiosis, and Healthy) disease classes. A key point in implementing transfer learning models is the process of freezing the first

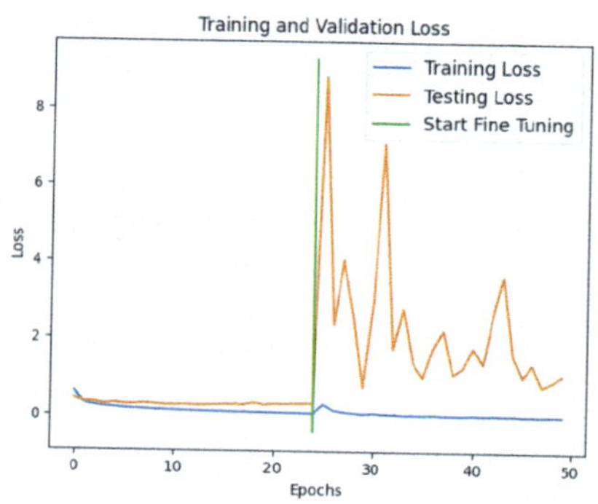

	precision	recall	f1-score	support
cocci	0.95	0.96	0.95	232
healthy	0.92	0.91	0.92	225
ncd	0.75	0.86	0.80	58
salmo	0.95	0.92	0.94	290
accuracy			0.93	805
macro avg	0.89	0.91	0.90	805
weighted avg	0.93	0.93	0.93	805

Accuracy score: 0.93
F1 score average: 0.9

Fig. 3 Results of training, testing, and prediction curves of the MobileNetV2 transfer learning model after hyperparameter tuning

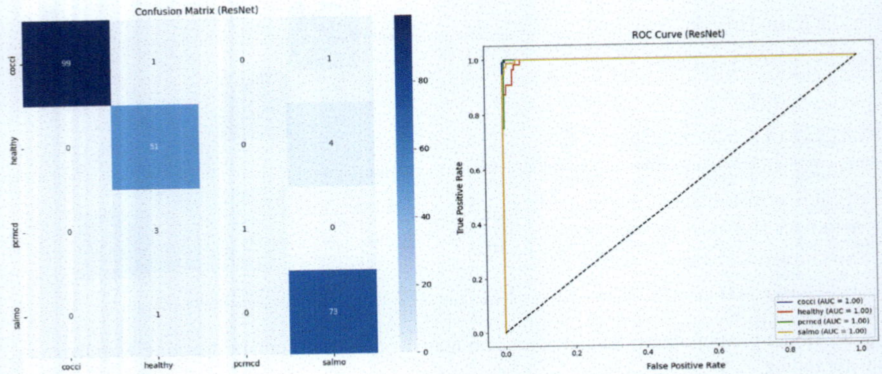

Fig. 4 Error matrices of models and ROC and AUC of the ResNet model

ResNet Report:

	precision	recall	f1-score	support
cocci	1.00	0.98	0.99	101
healthy	0.91	0.93	0.92	55
pcrncd	1.00	0.25	0.40	4
salmo	0.94	0.99	0.96	74
accuracy			0.96	234
macro avg	0.96	0.79	0.82	234
weighted avg	0.96	0.96	0.95	234

Fig. 5 Presented is a full report of the ResNet model for poultry disease prediction

Model : MobileNetV2 Model

	precision	recall	f1-score	support
cocci	0.95	0.96	0.95	232
healthy	0.92	0.91	0.92	225
ncd	0.75	0.86	0.80	58
salmo	0.95	0.92	0.94	290
accuracy			0.93	805
macro avg	0.89	0.91	0.90	805
weighted avg	0.93	0.93	0.93	805

Accuracy score: 0.93
F1 score average: 0.9

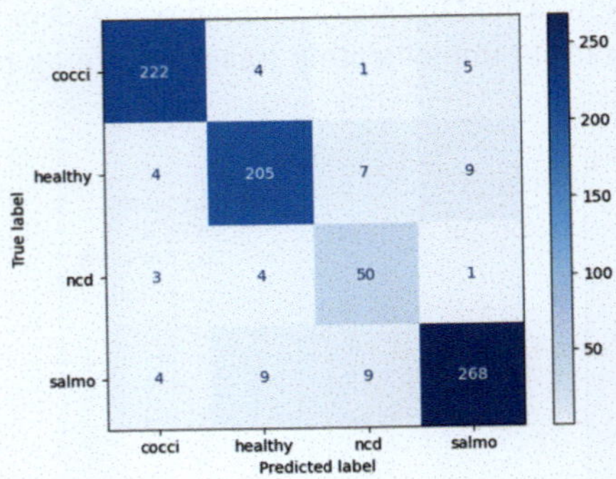

Fig. 6 MobileNetV2 model prediction report and error matrix results

layers of the neural network. When implementing this model, the first 150 layers of the neural network were frozen, and the training speed of the EfficientB4 model reached the power of 10^{-5}. The model was recompiled with a smaller learning rate. In this case, the model was further trained for up to 10 epochs. During implementation, this model showed the best results: Healthy 0.86, Newcastle 1, Coccidiosis 0.99, and Salmonellosis 0.96. The ROC-AUC curves were Newcastle: 0.99, Salmonellosis: 0.97, Coccidiosis: 0.98 and Healthy: 0.99.

Figures 7 and 8 show the additional training of the EfficientB4 model, the learning curves and the errors of the model before and after tuning the neural network hyperparameters.

The results of the EfficientB4 model's confusion matrix (see Fig. 9) show a noticeable improvement in the increase in positively predicted classes before and after tuning the network's hyperparameters. The predicted class columns of the confusion

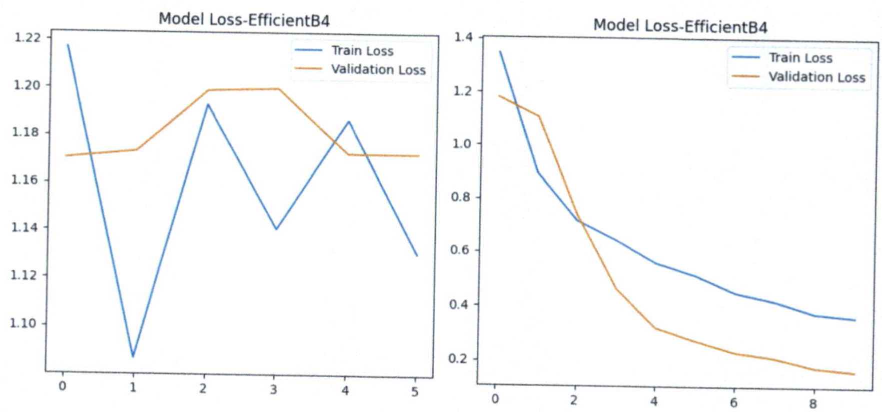

Fig. 7 Error and accuracy of the EfficientB4 model before hyperparameter tuning

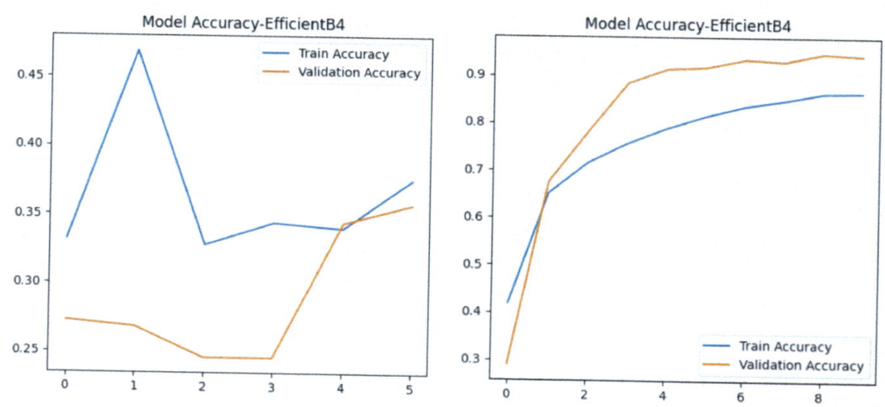

Fig. 8 Error and accuracy of the EfficientB4 model after hyperparameter tuning

matrix are a square table displaying the number of predicted and actual classes for the classification model. The model's predicted values are in the columns and represent the labels predicted by the model, while the rows of the matrix reflect the actual class labels (Fig. 10).

The following graph Fig. 11 shows the ROC AUC curves of the EfficientB4 model.

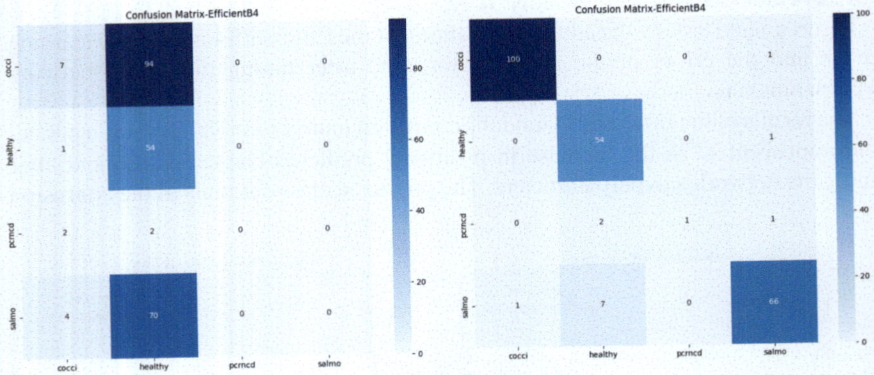

Fig. 9 Error matrix before and after hyperparameter tuning of the EfficientB4 model

	precision	recall	f1-score	support
cocci	0.50	0.07	0.12	101
healthy	0.25	0.98	0.39	55
pcrncd	0.00	0.00	0.00	4
salmo	0.00	0.00	0.00	74
accuracy			0.26	234
macro avg	0.19	0.26	0.13	234
weighted avg	0.27	0.26	0.14	234

	precision	recall	f1-score	support
cocci	0.99	0.99	0.99	101
healthy	0.86	0.98	0.92	55
pcrncd	1.00	0.25	0.40	4
salmo	0.96	0.89	0.92	74
accuracy			0.94	234
macro avg	0.95	0.78	0.81	234
weighted avg	0.95	0.94	0.94	234

Fig. 10 Full report of the forecast before and after hyperparameter tuning in the EfficientB4 model

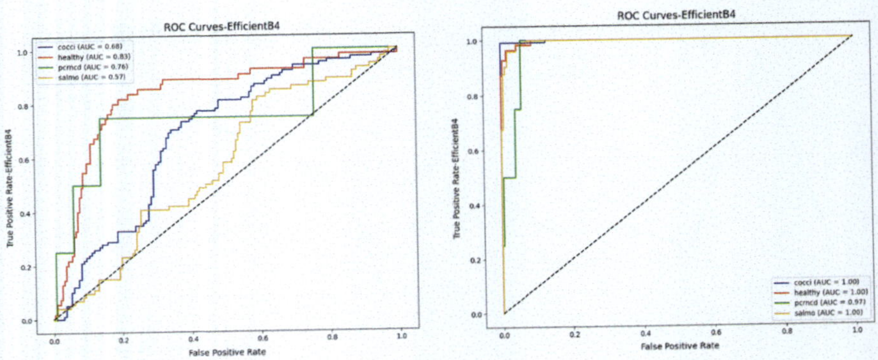

Fig. 11 Prediction report and ROC AUC curves before and after hyperparameter tuning in the EfficientB4 model

Table 2 Forecast results of the EfficientNetB4 model after fine-tuning

	Precision	Recall	f1-score	Support
Newcastle	1.00	0.9567	0.9617	4
Salmonellosis	0.9686	0.8901	0.9226	74
Coccidiosis	0.9912	0.9940	0.9926	101
Healthy	0.8667	0.9869	0.9268	55
Accuracy			0.9520	234
Macro avg	0.9533	0.9523	0.9528	234
Weighted avg	0.9520	0.9520	0.9520	234
Average AUC:0.983				

Table 1 below shows the training results of the EfficientNetB4 model (Table 2).

This implementation of the EfficientNetB4 model with the Adam optimizer achieved nearly 100% success rate and 99.9% accuracy on test data, making it robust and effective in predicting poultry diseases for automated avian disease diagnosis.

4 Conclusion

An important problem of the agricultural system for the poultry industry was studied. Using machine and deep learning technologies, models for recognizing poultry diseases were built. Neural modeling methods were used in constructing forecasting models. The constructed models generally demonstrated high results on test examples. The forecasting accuracy of the models was 98.45% when predicting salmonella. For example, various CNN variants demonstrated disease forecasting results ranging from 85 to 100% accuracy for certain poultry diseases. A model error matrix was constructed with the results of the largest number of correctly classified diseases, coccidiosis and salmonella. For the ResNet model, Fig. 4 shows the model error matrices, ROC, and AUC curves of the ResNet architecture for three diseases. The use of Adam optimizers was found to achieve good model performance in poultry farming, with an accuracy of 98.33% and a model error of 0.001. Using a halting neural network architecture or further training it using Keras gave us a noticeable increase in model accuracy. The data augmentation methods discussed above play a key role in allowing for slight overfitting in models, which immediately requires regularization methods. Depending on the tasks and the amount of data for model training, in most cases, the use of different approaches to model construction requires different neural network architectures, hyperparameter tuning, and the selection of epochs with training stops. The choice of epoch and neural network optimizers will yield different forecast results each time if overfitting occurs. In deep learning, with overfitting always present, the correct method for stopping model training depends on the choice of epoch. A key and important step in modeling agricultural problems is the formulation of the problem and the appropriate selection and tuning of hyperparameters, as well as the use of already trained models on big data, using transfer learning models as the primary deep learning method. For the MobileNetV2 model, after parameter tuning, the model accuracy for some diseases (coccidiosis) reached 95%. The confusion matrix and full report of the model prediction are shown in Fig. 6. The EfficientB4 prediction models with the Adam optimizer showed the best results, for example, Newcastle: 1. When reducing the learning rate and recompiling the model with freezing most of the neural network. Figure 11 shows the results of ROC and AUC curves for the EfficientB4 model, with the prediction quality results for the disease cocci (AUC = 0.68, AUC = 1)), salmo (AUC = 0.76, AUC = 0.97), healthy (AUC = 0.83, AUC = 1) and pcmcd (AUC = 0.76, AUC = 0.97). The prediction results of this model for the diseases cocci, salmo, healthy and pcmcd before 0.50, 0.00, 0.25, 0.00 after 0.99, 0.86, 0.96 and 1, respectively. The constructed model successfully detects poultry diseases with 98.3% accuracy. ROC curves confirm high predictive ability (AUC > 0.88 for all classes).

References

1. Neethirajan, S.: Automated tracking systems for the assessment of farmed poultry. Animals. **12**(3), 232 (2022). https://doi.org/10.3390/ani12030232
2. Conan, F.L.G.A., Sorn, S., Vong, S.: Biosecurity measures for backyard poultry in developing countries: a systematic review. BMC Vet. Res. **8**, 240 (2012)
3. Shapiro, B.I., Gebru, G., Desta, S., Negassa, A., Nigussie, K., Aboset, G., Mechal, H.: Ethiopia Livestock Master Plan. International Livestock Research Institute (ILRI), Nairobi, Kenya (2015)
4. Astill, J., Dara, R.A., Fraser, E., Sharif, S.: Detecting and predicting emerging disease in poultry with the implementation of new technologies and big data: a focus on avian influenza virus. Front. Vet.Y Sci. **5** (2018)
5. Ritchie, H., Rosado, P., Roser, M.: Meat and Dairy Production (2017). Published online at OurWorldInData.org. https://ourworldindata.org/meat-production
6. Xavier Averós, B.B., Cameno, E., Estevez, I.: The value of a retrospective analysis of slaughter records for the welfare of broiler chickens. Poult. Sci. (2020)
7. Bulent, T.E.K.E.: Survey on dead on arrival of broiler chickens under commercial transport conditions. Large Anim. Rev., 237–241 (2019)
8. Beretta, C., Stoessel, F., Baier, U., Hellweg, S.: Quantifying food losses and the potential for reduction in Switzerland. Waste Manag. **33**(3), 764–773 (2013). https://doi.org/10.1016/j.wasman.2012.11.007
9. Yohannes Asfaw, G.A., Medhin, G., Alemayehu, G., Wieland, B.: Publication Nfectious and Parasitic Diseases of Poultry in Ethiopia: A Systematic Review and Meta-Analysisauthor. Poultry Science Publisher (2019)
10. Asfaw, Z., Bihonegn, T.: Newcastle disease in Ethiopia: a review article. Int. J. Adv. Res. Biol. Sci. **5**(11), 95–102 (2018)
11. El-Saadony, M.T., Salem, H.M., El-Tahan, A.M., Abd El-Mageed, T.A., Soliman, S.M., Khafaga, A.F., Swelum, A.A., Ahmed, A.E., Alshammari, F.A., Abd El-Hack, M.E.: The control of poultry salmonellosis using organic agents: an updated overview. Poult. Sci. **101**(4), 101716 (2022). https://doi.org/10.1016/j.psj.2022.101716
12. Mesa-Pineda, C., Navarro-Ruíz, J.L., López-Osorio, S., Chaparro-Gutiérrez, J.J., Gómez-Osorio, L.M.: Chicken Coccidiosis: from the parasite lifecycle to control of the disease. Front. Vet. Sci. **21**(8), 787653 (2021). https://doi.org/10.3389/fvets.2021.787653

13. Getabalew, M., Alemneh, T., Akeberegn, D., Getahun, D., Zewdie, D.: Epidemiology, diagnosis & prevention of Newcastle disease in poultry. Am. J. Biomed. Sci. Res. **3**(1), 50–59 (2019). https://doi.org/10.34297/AJBSR.2019.03.000632
14. Sabitov, B., Asanbekova, N., Seitkazieva, N., Kartanova, A., Jentaeva, K.: Deep learning-based modeling and recognition of poultry diseases. In: Fourth International Conference on Optics, Computer Applications, and Materials Science (CMSD-IV 2024), Dushanbe, Tajikistan (2024) https://www.spiedigitallibrary.org/conference-proceedings-of-SPIE/13651.toc

Modeling Agricultural Problems Based on Neural Technologies

Ainura Dyikanova, Aida Turdubaeva, Turgunbek Zhumaliev, Sovetbek Mamytkanov, and Baratbek Sabitov

Abstract Models for predicting soil moisture were built based on agronomic and climatic data. Transfer learning models such as DenseNet, ResNet, and EfficientNet were used to identify complex nonlinear relationships between data. Improved model performance was achieved through regularization methods and tuning of model training hyperparameters. Smoother model learning curves were obtained, demonstrating the improved generalization ability of regularized models. To optimize the models, interactions between data features were analyzed. Correlations between features were constructed for model construction. Optimized models were constructed that take into account complex relationships in the data to build effective soil moisture forecasting models. A strong correlation was found between soil type and fertilizers. Climate factors such as temperature and air humidity were shown to interact nonlinearly. Nutrients such as N-nitrogen, P-phosphorus, and K-potassium exhibit complex interactions. The hybrid neural network models constructed with interaction blocks based on transfer learning and dense layers (Dense, ResNet, and EfficientNet) demonstrated the best results. Using a comprehensive analysis, neural network-based models were built with high performance compared to classical machine learning methods.

Keywords Soil moisture · Agricultural systems · Transfer learning · Deep learning · Neural technologies · Modeling · Forecasting · Model accuracy · Comparative analysis

---

A. Dyikanova · T. Zhumaliev · S. Mamytkanov
Kyrgyz National Agrarian University named after K.I. Scriabin, Bishkek, Kyrgyzstan

A. Turdubaeva
Kyrgyz Technical University named after I. Razzakova, Bishkek, Kyrgyzstan

B. Sabitov (✉)
Kyrgyz National University named after J. Balasagyn, Bishkek, Kyrgyzstan

D. Muhamediyeva et al. (eds.), *Engineering and Digital Technologies in Agriculture*, Smart Agriculture 20, https://doi.org/10.1007/978-981-92-0481-6_3

1 Introduction

Neural networks are a modern and powerful tool for modeling complex categories of agricultural systems, particularly soil moisture. Deep learning tools used enable the construction of complex nonlinear relationships between input features and soil moisture for model building. Analyzing soil water movement and scheduling irrigation are crucial for agricultural irrigation of many crops. Furthermore, predicting the quantitative estimate of water use and its assessment forms the basis of water resource management in agriculture. This study examines machine learning and deep learning neural network methods for building validated soil moisture forecasting models, a key task in precision farming. Deep learning methods and the closely related neural networks offer powerful tools for predicting soil moisture by analyzing various relationships in data and complex categories of feature space. These methods can be used to evaluate models for predicting soil moisture using a wide range of data, including satellite imagery, large-scale ground-based instrumentation, and even visual characteristics such as color and texture. By training deep learning models on large datasets, it is possible to accurately predict soil moisture levels, which is critical for precision agriculture and crop water management. Soil moisture sensors are used in numerous research applications, such as agricultural applications and fruit crop production, including irrigation management, depending on changing climate conditions. Soil health is a complex aspect of agriculture and is considered a key criterion for implementing a food program that optimizes natural resources [1]. Soil moisture is an important factor for food security, water management, and agricultural irrigation. Another important factor for soil is solar energy received from the atmosphere [2]. The amount of water contained in the soil and its consumption determine soil moisture, which varies over time, depending on the water consumption of various types of plants, losses in agricultural land, and its aridity. Key factors in maintaining soil moisture include a variety of key climatic parameters, including the amount of precipitation in agricultural land, the type of field irrigation, the type of crops grown, the seasonality of water consumption in the soil, and natural evaporation [3]. Due to climate change, abnormal temperature events, heavy precipitation, and other climatic factors lead to fluctuations in soil moisture values [4]. To determine the health and productivity of agricultural crops, the main task of farmers is to optimize and manage soil moisture and irrigation intensity [2]. It is important to use modern tools that have a greater capacity for continuous monitoring and forecasting of soil moisture. A major disadvantage of farmers is the use of laboratory analysis to forecast soil moisture [5–8]. However, this is time-consuming, expensive, and inaccessible to many farmers. These expensive laboratory and analytical methods are costly for many farmers and are short-term, requiring ongoing financial investment [8, 9]. Using data from a large area, remote sensing methods for predicting soil moisture can provide farmers with positive results. However, this method is only beneficial for farmers with large agricultural lands [9]. In recent years, with the development of artificial intelligence tools, deep learning technologies have been used to achieve the desired accuracy and build models for predicting soil moisture. Soil color data is very important when using deep learning methods, which can classify soil moisture from images. Neural networks with various architectures are among the modern research methods where modeling using neural technologies can be used [5, 7, 10, 11].

Using a feature space constructed from various image data structures, such as color, numerical, and categorical soil values, or historical soil moisture data for specific agricultural fields, machine learning algorithms can be used to predict soil moisture with any degree of accuracy. To predict the objective function that defines the task and to establish relationships between features that characterize the soil, machine learning methods can be used to extract patterns and feature knowledge, which is essential for forecasting using other methods. Artificial intelligence methods require the collection of large data sets of various types to construct moisture models, which can be accomplished using manual assembly or hardware devices such as smartphones or drones with specially equipped high-resolution video cameras for collecting soil images [6].

Modern computer vision methods with deep learning technologies and convolutional neural networks (CNN) based on big image data can ensure rapid and accurate determination of soil characteristics. For forecasting major problems with irrigation of agricultural land for various types of cultivated plants, forecasting using soil images allows the creation of highly accurate models. For the initial analysis of models, linear models can be constructed, including regression and logical models of soil moisture, which identify linear patterns in characteristics [12]. To construct and forecast time-dependent models based on climatic, agricultural, and soil data for regional conditions, linear regression models based on machine learning algorithms can be built, for example, for weekly irrigation requirements for agricultural land. Moreover, the construction of regional models taking into account agricultural and climatic parameters can have a significant effect for individual cultivated crops [13].

Taking into account local meteorological conditions is a crucial requirement for building models for forecasting water consumption in agricultural lands for decision-making in building universal models. In [14], a soil moisture deficit model based on climate and soil moisture data is considered. To establish a link between the input data and the output space, a comparative analysis [15] showed that machine learning methods show higher performance than classical statistical methods. However, statistical methods using spatial and temporal patterns hidden in historical data can produce the desired results and models. A feature of this approach is not to rely on a physical model of the system, but on advanced statistical data. For forecasting tasks, for example, a complex category of soil moisture, it is desirable to have models built taking into account the dynamics of the data [16]. However, building such models requires physical factors, such as soil characteristics for a given region or susceptibility to erosion and data on advanced soil properties, which complicates the construction of models based on traditional machine learning methods. This is due to the complex nonlinear relationship between climatic parameters, soil hydraulic properties and soil moisture dynamics [17]. Modeling processes that require more sophisticated artificial intelligence tools, such as neural modeling with computer vision technologies, is essential. Computer vision technologies with deep learning methods can produce highly accurate, decisive models in this case. However, this requires extensive image data on soil moisture depending on regional meteorological conditions, as well as data on agronomic soil properties and climate change factors, which complicates data collection. However, for this case, there are preliminary models of experimental value [6, 11] that use existing machine learning methods to estimate soil moisture based on images obtained with modern technical means. Constructing models using soil images is effective when transfer learning methods are used. However, in this case, collecting a large number of soil images is difficult and labor-intensive. Approaches based on machine learning algorithms can be used on numerical data; categorical data can yield good results in comparison.

Results obtained using DenseNet transfer models with dense layers of such data yield high-quality models for predicting soil moisture. In this case, a comparative analysis of machine learning methods with transfer learning can be used on multi-modal data of various types. It is important to emphasize the accuracy and quality of images collected in the field, which play a significant role in building high-quality models. For collecting information, it is crucial to have a mobile app or other technical means for collecting soil data that can be accessed by a wide audience of farmers and agricultural producers, allowing them to quickly, affordably, and effectively estimate soil moisture in their agricultural fields in the future. Examples of such web-based systems include systems using centralized database servers, such as various versions of PostGreeSQL, with a web interface in Django. It is also important to consider the role of satellite soil data, which could fundamentally change data collection technologies. The main goal is to develop effective and accessible models for predicting soil moisture and other criteria for improving soil productivity.

2 Materials and Methods

Machine learning algorithms are used in this study to construct classical models for predicting soil moisture. Deep learning methods are employed to identify complex, nonlinear interactions between features. A key area of research is pre-trained transfer learning models. Below, we present a mathematical description of the main machine learning algorithms, deep learning methods, and the application of transfer learning to a class of soil moisture forecasting problems. Transfer learning models are pre-trained on big data, making them convenient for training new models while freezing the weights of previously trained models. Representatives of this class of pre-trained models, such as DenseNet and ResNet, are used in this study. Below, we present a formal description of key machine learning (ML) and deep learning (DL) algorithms, including their objective functions, optimization methods, and architectures. Using an objective function as the result of linear regression for initial modeling is the initial step toward modeling complex processes. Predicting a continuous variable y based on a linear combination of features x using the model $y = w^T x + b$, where w is the weight and b is the bias, allows us to estimate the initial loss function MSE:

$$\mathcal{L}(w, b) = \frac{1}{N}\sum_{i=1}^{N}(y_i - y)^2 \tag{1}$$

The optimization of the model is based on gradient descent, the criterion for determining the parameters of which is based on the optimization problem:

$$\mathcal{L}(w, b) = \frac{1}{N}\sum_{i=1}^{N}\left(y_i - w^T x_i - b\right)^2 \rightarrow \min \tag{2}$$

Unknown parameters of the model are updated by recurrence relations:

$$w \leftarrow w - \eta \nabla_w \mathcal{L}, \ \ b \leftarrow b - \eta \frac{\partial}{\partial b}\mathcal{L} \tag{3}$$

Equation (2) has a direct analytical exact solution (4) of problem (2), called the normal solution. The general form of this solution in matrix notation is:

$$w = \left(X^T X\right)^{-1} X^T y \tag{4}$$

An important algorithm for constructing models for classification problems is logistic regression. The main task of this algorithm is to classify a target binary function with an output in the form of a probability $P(y = 1|x)$. In this case, the activation function (sigmoid) is described as:

$$P(y = 1|x) = \sigma\left(\boldsymbol{\omega}^T x + b\right), \quad \sigma(z) = \frac{1}{1 + e^z}. \tag{5}$$

The loss function is defined using Log Loss, which is determined by the quality functional:

$$\mathcal{L}(\boldsymbol{\omega}, b) = -\frac{1}{N}\sum_{i=1}^{N}\left[\, y_i log P(y_i|x_i) + (1 - y_i)\log(1 - P(y_i|x_i))\,\right], \tag{6}$$

For the problem of minimizing the loss function, gradient descent is used, the parameters of which are updated according to the formula.

$$\boldsymbol{\omega} \leftarrow \boldsymbol{\omega} - \eta \sum_{i=1}^{N}(P(y_i|\boldsymbol{x}_i) - y_i)\boldsymbol{x}_i, \tag{7}$$

The next important algorithm for data analysis is the advanced support vector machine (SVM), which aims to maximize the distance between hyperplanes, i.e., the gap between classes. The hyperplane is described by the equation:

$$w^T x + b = 0 \tag{8}$$

The optimization problem for this method is given by the general formula:

$$\min_{w,b} \frac{1}{2}\|w\|^2 + C\sum_{i=1}^{N} \max\left(0, 1 - y_i\left(\boldsymbol{w}^T \boldsymbol{x}_i + b\right)\right) \tag{9}$$

where C is the regularization parameter. Algorithms that produce high prediction rates include gradient boosting (XGBoost, LightGBM). It is typically used to achieve high accuracy results. The minimization problem with a loss function with regularization for this method is defined as:

$$\mathcal{L} = \sum_{i=1}^{N} L(y_i, \hat{y}_i) + \sum_{m=1}^{M} \Omega(f_m) \rightarrow \min, \quad \Omega(f) = \gamma T + \frac{1}{2}\lambda||w||^2 \tag{10}$$

where $L(y_i, \hat{y}_i)$ is the quality functional. We will now describe some key aspects of the mathematical foundation of neural networks and deep learning. One of the key methods for initially constructing a fully connected neural network, or MLP for short, is a connected feedforward network. The formula used to transfer data between neurons is

$$z^{|l|} = W^{|l|}a^{|l-1|} + b, \quad a^{|l|} = \sigma\left(z^{|l|}\right) \tag{11}$$

where σ is the activation function that introduces nonlinearity into the model, and it defines the output signal using the weighted sum of its inputs in the form (11) with vector weighting coefficients $W^{(l)}$. The activation functions for each neuron are specified as functions (12)–(14). Examples of activation functions used in neural networks are given in (12)–(14).

$$\text{ReLU} : \sigma(z) = \max(0, z) \tag{12}$$

$$\text{Sigmoid} : \sigma(z) = \frac{1}{1 + e^z} \tag{13}$$

$$\text{Softmax (for classification)} : \sigma(x_i) = \frac{e^{x_i}}{\sum_{j=1}^{N} e^{x_j}}. \tag{14}$$

Backpropagation using convolutional networks (CNNs) uses the mathematical concept of convolution. Convolution is an operation applied to input data (images, signals) using a filter (kernel). The formula for discrete convolution for a 2D image is defined as follows:

$$\left(S^* K\right)(i, j) = \sum_m \sum_n S(i - m, j - n) \cdot K(m, n) \tag{15}$$

where S is the input matrix (image), K is the convolution kernel (filter), and (i, j) are the coordinates of the output value.

Pooling is a dimensionality reduction operation that preserves important features. Max Pooling:

$$\text{MaxPool }(A)_{i,j} = \max_{m,n\in\mathcal{R}} A_{i+m,j+n} \tag{16}$$

where $A_{(i,j)}$ are image elements. Various algorithms are used for optimization in neural networks, for example, stochastic gradient descent (SGD) is defined by a recurrent formula.

$$w \leftarrow w - \eta\nabla_w\mathcal{L}(w), \tag{17}$$

where η is the learning rate of the method. Adam (short for "adaptive momentum estimation") is an optimization algorithm that combines ideas from SGD with momentum and RMSProp. This algorithm is a fundamental algorithm and is used in many deep learning architectures. Similar to RMSProp, it uses an exponentially decaying average of gradients and squared gradients to determine the updated scale. It also uses momentum to help the optimizer navigate the loss function more efficiently. The algorithm for updating the iteration over the variables w_t can be written as follows:

$$m_t = \beta_1 m_{t-1} + (1 - \beta_1)\nabla_w\mathcal{L}(w_t), \tag{18}$$

$$v_t = \beta_2 v_{t-1} + (1 - \beta_2)\left(\nabla_w\mathcal{L}(w_t)\right)^2 \tag{19}$$

$$w_{t+1} = w_t - \eta\frac{m_t}{\varepsilon + \sqrt{v_t},} \tag{20}$$

where m and v are the momentum and velocity vectors, respectively, and β_1 and β_2 are the decay rates of the momentum and velocity. This method, used in this paper, calculates the adaptive learning rate for each parameter. It stores both the decaying mean of past gradients, analogous to momentum, and the decaying mean of past squared gradients, similar to RMSProp. Thus, it combines the advantages of both methods. This algorithm converges faster than other optimization algorithms and performs well on noisy data.

To evaluate the model performance and their effectiveness, the following accuracy metrics were applied to the trained soil moisture forecasting models. The accuracy of the model between the observed and predicted soil moisture was assessed using the coefficient of determination (R^2), the root mean square error (RMSE), and the mean absolute error (MAE). The R^2 score of a model describes the proportion of the total variance in the observed data that is explained by the model and varies within the range $[-\infty, 1]$. If the R^2 value of the constructed model is close to 1, then the score indicates that the model explains well the squared deviation from the mean, i.e., the variance of the observations. The formula for calculating R^2 is determined by formula (21):

$$R^2 = \frac{\sum_{i=1}^{N} (y_i - \overline{y})^2 - \sum_{i=1}^{N} (y_i - \hat{y}_i)^2}{\sum_{i=1}^{N} (y_i - \overline{y})^2} \quad (21)$$

The MSE and MAE quantify forecast errors in the same units of measurement for the variables. However, the RMSE heavily penalizes large outliers; therefore, it is preferable to supplement it with the MAE and MSE. Values close to zero indicate good model predictions. The mean square error (MAE) and root mean square error (RMSE) are defined by the following formulas:

$$\text{MAE} = \frac{1}{n}\sum_{i=1}^{n} |y_i - \hat{y}_i| \quad (22)$$

$$\text{RMSE} = \left[\frac{1}{n}\sum_{i=1}^{N} (y_i - \hat{y}_i)^2\right]^{0,5} \quad (23)$$

In Eqs. (21), (22) and (23) y_i are observations and $\hat{y}_i$ is the predicted value at time i ($i = 1, 2, \ldots, n$).

3 Results and Discussion

Pairwise interactions between traits were studied, and optimization models were constructed. To build more effective soil moisture forecasting models, interactions between traits were thoroughly analyzed. This enabled the creation of optimization models that account for complex relationships in the data. An analysis of pairwise interactions between traits with hidden relationships was obtained. Figures 1 and 2 present the correlation and regression models as a graph of the relationship between temperature, atmospheric humidity, and the agronomic nutrient parameters nitrogen, phosphorus, and potassium.

In Fig. 3 the results of the distribution of the interaction "Temperature and Air Humidity. Soil Moisture and Nutrients" are shown.

Figure 4 shows the linear regression model for predicting soil moisture and the mutual correlations of soil type, nutrients and soil moisture (see Fig. 4a) and the regression model of soil moisture (see Fig. 4b).

Figure 5 the root mean square error MSE for estimating the error of the DenseNet model for predicting soil moisture and the model built with regularization b) were obtained.

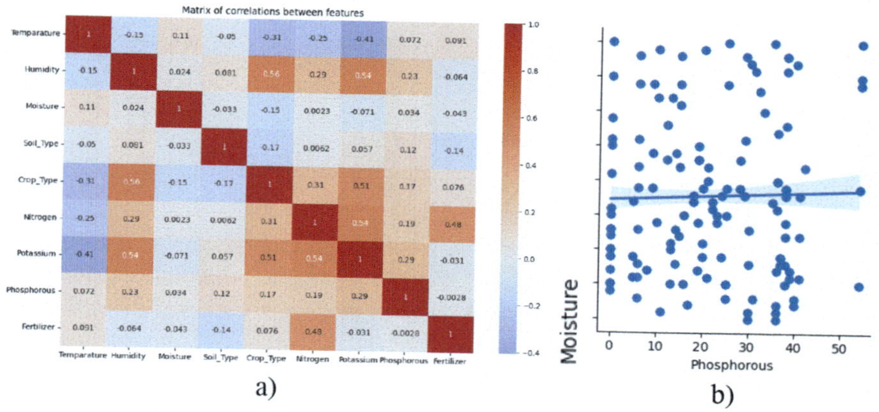

Fig. 1 Correlation matrix of the complex relationship between the features and the linear dependence in the form of a regression relationship between soil moisture and temperature

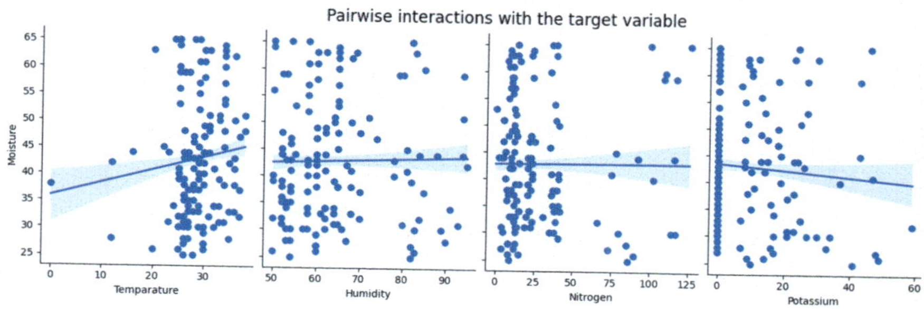

Fig. 2 Pairwise interactions of soil characteristic traits with the target variable soil moisture

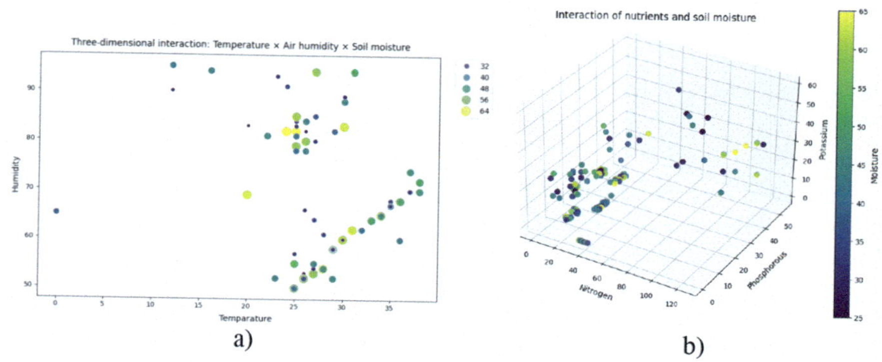

Fig. 3 Interaction of air temperature and humidity and soil moisture "Temperature—Air Humidity and Nutrients"

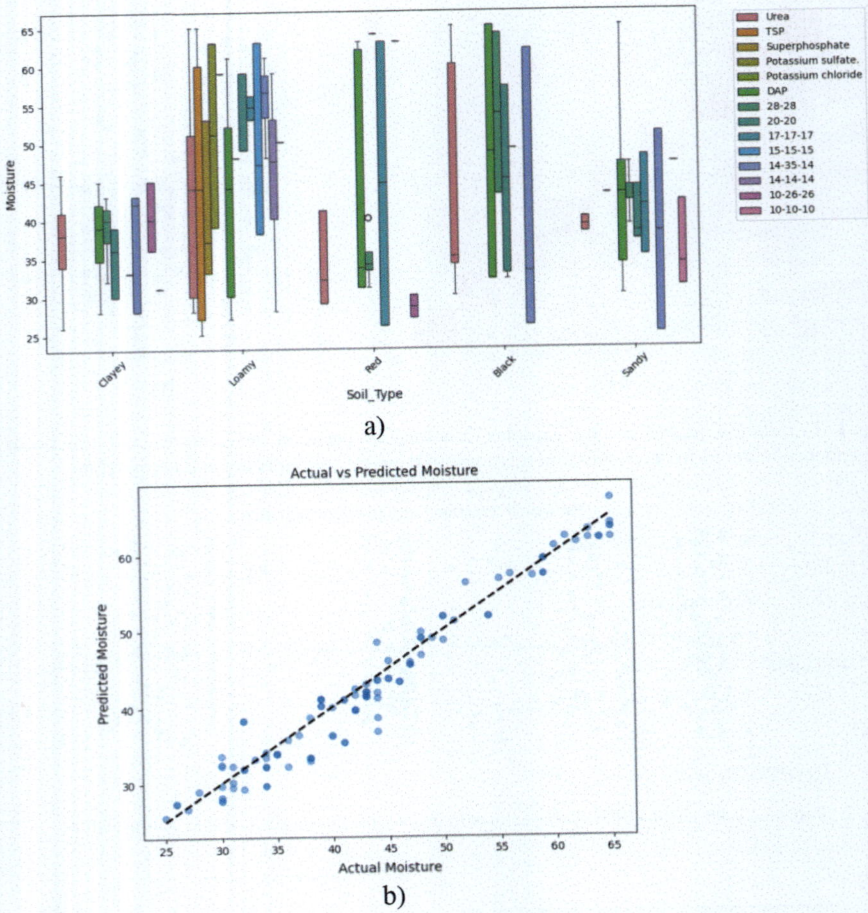

Fig. 4 Cross-correlations between soil types, nutrients and a linear regression model of soil moisture

Figure 6 shows the results of the mean absolute error -MAE using the implementation of the DenseNet neural model and the regularized neural model and their dependence on the training epochs for predicting soil moisture.

The results of the analysis of the DenseNet and Resnet models for soil moisture forecasting and their comparative model in various MSE and MAE metrics, as well as classical machine learning regression models with the DenseNet and Resnet neural networks, were obtained. For the preliminary analysis, it was also important to evaluate the model in various metrics using machine learning methods. Listing 2 shows a comparison of metric evaluations using machine learning algorithms: gradient boosting, random forest, Lasso and Ridge regression, linear regression,

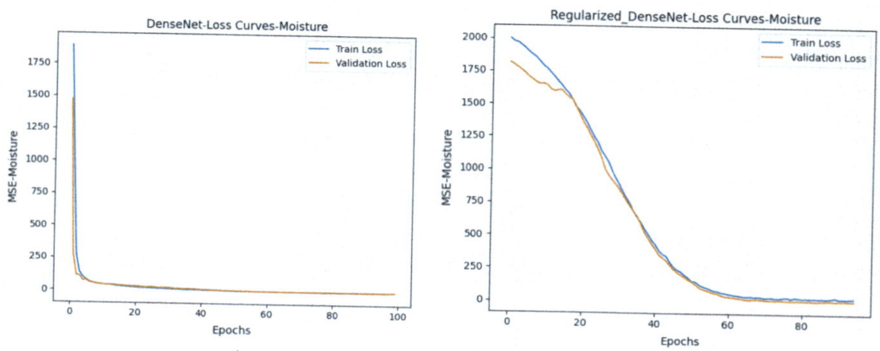

a) b)

Fig. 5 Distribution of MSE soil moisture depending on epochs and regularization

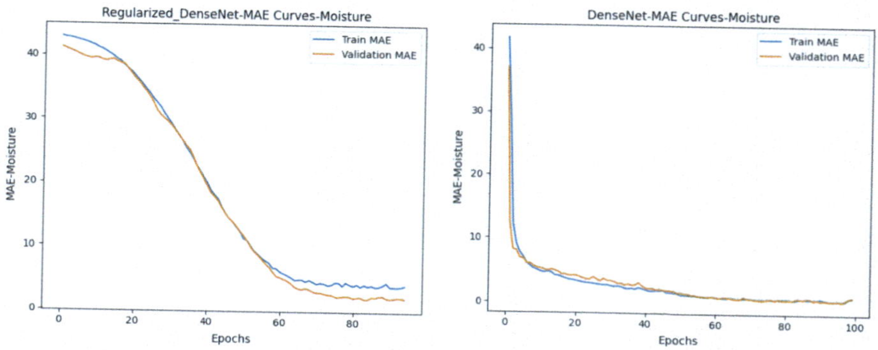

a) b)

Fig. 6 Error curves for comparison of absolute error -MAE and soil moisture depending on regularization and epochs

support vector machines, XGBoost, nearest neighbor matching, and transfer learning methods.

Listing 1. Comparison of metrics of machine learning methods and neural networks.

```
Final comparative analysis of models:
                                 MSE        MAE        R2
Gradient Boosting         146.472566   9.944130  0.074247
Random Forest             149.149938   9.871726  0.057325
ResNet                    154.099342  10.219800  0.026044
Lasso Regression          154.548792  10.317369  0.023203
Ridge Regression          155.013195  10.313943  0.020268
Linear Regression         155.015054  10.313949  0.020256
Support Vector Regression 155.958448  10.088648  0.014293
XGBoost                   169.532660  10.492411 -0.071500
K-Nearest Neighbors       183.556914  10.952523 -0.160138

Best model: Gradient Boosting c R2 = 0.0742

ResNet improves the best classical model by-54.57% по R2
```

Models of training error curves, the result of Fig. 7, the implementation of the conventional ResNet model and the regularized model for predicting soil moisture using the example of the ResNet transfer learning model on training and validation data sets.

In Fig. 8 is a graph of regularized models for comparative analysis of the DenseNet and ResNet regression models for soil moisture forecasting.

For the DenseNet model, the performance of various model metrics before and after the regularization method had the following values: Listing 2.

Listing 2. DenseNet model metric results:

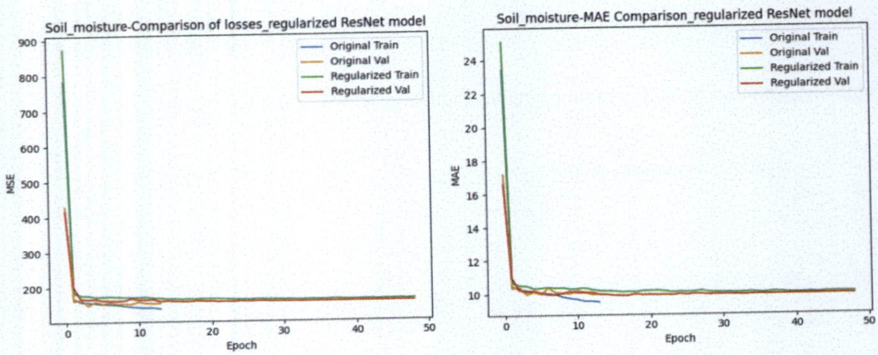

Fig. 7 Distribution of MSE and MAE of soil moisture depending on the epochs

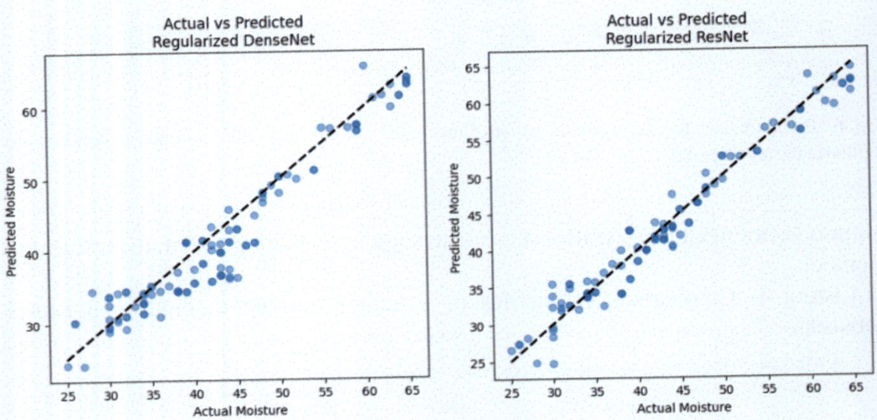

Fig. 8 Prediction of soil moisture by linear regression model depending on epochs

```
                 Model       MSE      RMSE       MAE  R2 Score  \
0         Base DenseNet  3.557002  1.886002  0.992612  0.968382
0  Regularized DenseNet  5.890467  2.427028  1.861294  0.947641

   Mean Absolute Percentage Error
0                        2.449423
0                        4.634455

Residuals Analysis:
Mean of residuals: 0.7502
Standard deviation of residuals: 2.3082
Minimum residual: -6.2592
Maximum residual: 7.2451
```

A comprehensive analysis of plant dependence on soil moisture and acidity, nutrients, and climatic parameters was conducted. Listing 3.

Listing 3. Plant requirements in relation to agricultural and climatic parameters.

```
Some Interesting Patterns:
--------------------------
Crops which requires very High Ratio of Nirtogen Content in Soil: ['cotton']
Crops which requires very High Ratio of Phosphorous Content in Soil: ['grapes' 'apple']
Crops which requires very High Ratio of Potassium Content in Soil: ['grapes' 'apple']
Crops which requires very High Rainfall: ['barley' 'apple']
Crops which requires very Low Temperature: ['barley' 'maize' 'potato' 'alfalfa' 'pear' 'cherry' 'mungbean'
 'blackgram' 'pomegranate' 'raspberry' 'corn' 'grapes' 'watermelon'
 'muskmelon' 'apple' 'cotton']
Crops which requires very High Temperature: ['grapes' 'barley']
Crops which requires very Low Humidty: ['barley' 'maize' 'alfalfa' 'pear' 'cherry' 'mungbean' 'blackgram'
 'pomegranate' 'raspberry' 'corn' 'grapes' 'watermelon' 'muskmelon'
 'apple' 'potato' 'cotton']
Crops which requires very Low pH: ['cherry']
Cr0ps which requires very Low pH: ['cherry']
```

4 Conclusion

An analysis of machine learning methods and various neural network architectures for predicting soil moisture yielded the following conclusions. Among classical models, the ensemble methods XGBoost, Gradient Boosting, and Random Forest demonstrated the best performance. The ResNet neural network model demonstrated comparable or better performance compared to classical machine learning methods. Transfer learning expanded the ResNet architecture analytics by improving the residual block architecture with a gradual dimensionality reduction according to the following Scheme (256 → 128 → 64). Linear model construction methods (linear regression, Ridge, and Lasso) showed worse results, indicating the nonlinear nature of the data. The model quality indicator in the R^2 quality metric (the coefficient of determination) is closer to 1 for the best constructed models. In these metrics, the MSE and MAE values are significantly lower for ensemble methods and ResNet. Using the XGBoost or Gradient Boosting algorithms, a mutual balance between accuracy and model training speed was achieved. ResNet achieved maximum accuracy in predicting soil moisture. However, this model requires more computational resources during training. To ensure interpretability, feature importance was analyzed using the Random Forest algorithm. One of the key findings in the analysis of the baseline models was the following shortcomings. When training the DenseNet model, learning curves and model errors overfitted, i.e., a gap was observed between the train and validation losses. The regularized model with Dropout, BatchNorm, and L_2 regularization demonstrated better generalization ability. Early stopping during model training helped optimize training time. Among the model quality metrics, the regularized model has a higher R^2 score and smaller MAE and MSE errors. Furthermore, the error distribution obtained was close to normal with low variance. A comparison of transfer learning neural network architectures revealed the following results. The ResNet transfer learning model showed slightly better results compared to the DenseNet model with dense layers due to residual connections.

Both architectures were susceptible to overfitting during training without regularization. The effectiveness of regularization methods was demonstrated in the following cases: L_1 regularization resets many weights to zero; this method works with features with weights close to zero. In this case, this approach, L_2 regularization, preserves important neural connections and selects useful features. It identifies many correlated features while discarding unimportant features. The third type of ElasticNet regularization, a combination of L_1/L_2 regularization and Dropout, significantly reduced overfitting of models with the identification of large correlations. A major advantage of regularized models is smoother learning curves and better generalization ability. The best result, with an $R^2 = 0.92$, was obtained for the regularized ResNet model. The MAE metric of the best model was 0.45 moisture units. To improve the quality of the models, additional hyperparameters were tuned for the model learning rate, the number of layers/neurons, and epoch management. Residual analysis revealed that the model underperforms in predicting extreme moisture values. The following features were observed in constructing the correlation matrix. A strong interaction was found between soil type and fertilizer. Climate factors (temperature and air humidity) interact nonlinearly. Nutrients (N-nitrogen, P-phosphorus, and K-potassium) exhibit complex interactions. Optimization models with explicit polynomial interactions improved the MSE by 15% compared to the baseline model. The model predicts soil moisture fairly well based on the provided features. Extreme moisture values contribute the most to the error. Temperature and soil type, according to residual analysis, are the most important features. The baseline DenseNet model showed signs of overfitting: there is a gap between the learning curves on the training and validation data.

A hybrid neural network model with interaction blocks demonstrated the best results. The optimization algorithm enables finding optimal parameters for specific conditions. It was found that for loamy soils, the optimal nitrogen level is above average. A combination of potassium and phosphorus fertilizers produces a synergistic effect. Under high temperatures, frequent, low-dose irrigation is more effective. The depth of moisture must also be taken into account. Overall, this analysis allows not only to predict soil moisture but also to actively manage it, selecting the optimal combination of factors for specific agricultural conditions.

References

1. Chatterjee, S., Dey, N., Sen, S.: Soil moisture quantity prediction using optimized neural supported model for sustainable agricultural applications. Sustain. Comput. Inform. Syst. **28**, 100279 (2020)
2. Pekel, E.: Estimation of soil moisture using decision tree regression. Theor. Appl. Climatol. **139**, 1111–1119 (2020)
3. Prakash, S., Sekhar, S.: Soil moisture prediction using shallow neural network. Int. J. Adv. Res. Eng. Technol. **11**, 426–435 (2020)
4. Han, J., Mao, K., Xu, T., Guo, J., Zuo, Z., Gao, C.: A soil moisture estimation framework based on the CART algorithm and its application in China. J. Hydrol. **563**, 65–75 (2018)
5. Zanetti, S.S., Cecílio, R.A., Alves, E.G., Silva, V.H., Sousa, E.F.: Estimation of the moisture content of tropical soils using colour images and artificial neural networks. Catena. **135**, 100–106 (2015)
6. Aitkenhead, M.J., Poggio, L., Wardell-Johnson, D., Coull, M.C., Rivington, M., Black, H.I.J., Yacob, G., Boke, S., Habte, M.: Estimating soil properties from smartphone imagery in Ethiopia. Comput. Electron. Agric. **171**, 105322 (2020)
7. Dos Santos, J.F.C., Silva, H.R.F., Pinto, F.A.C., De Assis, I.R.: Use of digital images to estimate soil moisture. Rev. Bras. Eng. Agric. E Ambient. **20**, 1051–1056 (2016)
8. Saad Hajjar, C., Hajjar, C., Esta, M., Ghorra Chamoun, Y.: Machine learning methods for soil moisture prediction in vineyards using digital images. In: Proceedings of the E3S web of conferences: 2020 11th international conference on environmental science and development (ICESD 2020), vol. 167, p. 7. Barcelona, Spain, 10–12 EDP Sciences: Les Ulis, France (2020)
9. Zheng, L., Li, M., Sun, J., Zhang, X., Zhao, P.: Estimating soil moisture based on image processing technologies. In: Applications of digital image processing XXVIII, vol. 5909, pp. 548–555. SPIE: Bellingham, WA, USA (2005)
10. Persson, M.: Estimating surface soil moisture from soil color using image analysis. Vadose Zone J. **4**, 1119–1122 (2005)
11. Taneja, P., Vasava, H.K., Daggupati, P., Biswas, A.: Multi-algorithm comparison to predict soil organic matter and soil moisture content from cell phone images. Geoderma. **385**, 114863 (2021)
12. Navarro-Hellín, H., Martínez-del-Rincon, J., Domingo-Miguel, R., Soto-Valles, F., Torres-Sánchez, R.: A decision support system for managing irrigation in agriculture. Comput. Electron. Agric. **124**, 121–131 (2016)
13. Giusti, E., Marsili-Libelli, S.: A fuzzy decision support system for irrigation and water conservation in agriculture. Environ. Model. Softw. **63**, 73–86 (2015)
14. Delgoda, D., Saleem, S.K., Malano, H., Halgamuge, M.N.: Root zone soil moisture prediction models based on system identification: formulation of the theory and validation using field and AQUACROP data. Agric. Water Manag. **163**, 344–353 (2014)
15. Karandish, F., Šimůnek, J.: A comparison of numerical and machine-learning modeling of soil water content with limited input data. J. Hydrol. **543**, 892–909 (2016)
16. McCarthy, A.C., Hancock, N.H., Raine, S.R.: Advanced process control of irrigation: the current state and an analysis to aid future development. Irrig. Sci. **31**, 183–192 (2013)
17. Mashayekhi, P., Ghorbani-Dashtaki, S., Mosaddeghi, M.R., Shirani, H., Nodoushan, A.R.M.: Different scenarios for inverse estimation of soil hydraulic parameters from double-ring infiltrometer data using HYDRUS-2D/3D. Int. Agrophys. **30**, 203–210 (2016)

Reverse Engineering of Agricultural Machinery Components Using Combined 3D Scanning and Parametric Modeling Technologies

Maria Karelina, Maxim Pletnev, and Dmitriy Taldykin

Abstract This article proposes a comprehensive methodology for reverse engineering components of agricultural machinery. The methodology uses a mixed approach combining 3D scanning and parametric modeling. The present study is devoted to the use of the Scanform L5 portable laser scanner, which uses laser triangulation to obtain spatial digitization of complex shaped parts with an average accuracy of $\pm$ 0.05 mm and a spatial resolution of up to 0.14 mm. Preliminary laboratory studies have confirmed the effectiveness of the scanner in digitizing the external surfaces of agricultural machinery parts for which there is no design documentation. The data obtained from the point cloud is processed using the Geomagic Design X post-processing software and imported into CAD systems, which allows you to create editable solid-state models. This workflow facilitates rapid reproduction, retrofitting, or re-fabrication of key components such as the spray nozzle brackets. The proposed approach is designed to solve such production problems as the need to modernize existing technological equipment, eliminate the shortage of spare parts and technological import substitution. This approach offers a reliable and scalable solution for ensuring operational availability and extending the service life of agricultural machinery.

Keywords Reverse engineering · Agricultural machinery · 3D scanning · Parametric modeling · Import substitution

1 Introduction

In the context of the growing demands for technological independence in the agricultural sector, reverse engineering has become a sought-after tool for developing design documentation and manufacturing important components of machine systems. The deterioration of the characteristics of the equipment in use in the case of untimely and

M. Karelina · M. Pletnev · D. Taldykin (✉)
State University of Management, Moscow, Russia
e-mail: ds_taldykin@guu.ru

D. Muhamediyeva et al. (eds.), *Engineering and Digital Technologies in Agriculture*, Smart Agriculture 20, https://doi.org/10.1007/978-981-92-0481-6_4

substandard maintenance, the shortage of components and dependence on foreign manufacturers in agricultural machinery are urgent problems.

This article suggests an approach, the implementation of which will solve the identified problems by reverse engineering critical components of agricultural machinery through structured light scanning and grid transformation. This approach has been developed taking into account the specific functional requirements of agricultural machinery.

Modern agricultural machinery often operates in conditions characterized by fluctuations in mechanical loads (for example, from 1.5 to 3.8 MPa on tillage units), the interaction of abrasive particles and frequent thermal cycles. Consequently, components such as spray gun brackets, seed dispenser shafts, and gear housings are subject to non-linear wear and failures. In the course of field research, it was found that 27.5% of equipment downtime was due to the lack of spare parts, while the average time to replace imported parts was 36–42 days. These figures demonstrate the importance of ensuring technical readiness and minimizing machine downtime.

Unlike classical modeling in a computer-aided design system, reverse engineering uses a reverse workflow in which the physical geometry of a part, often without documentation, is transformed into a parametric or mesh digital model. The effectiveness of the process depends on the use of metrological survey methods used, which are characterized by the ability to achieve a high level of resolution [1, 2].

The continuous development of engineering analysis and industrial quality control has highlighted the need for accurate, fast and contactless methods for digitizing geometric data, especially in the context of reverse engineering, when reproducing physical objects must be accurate to the original surface geometry.

Handheld 3D scanners (see Fig. 1) are designed as high-precision metrological devices designed to digitize the geometry of the studied objects using the principles of laser triangulation in combination with advanced photogrammetric algorithms. The system meets the urgent need for portable and flexible measuring solutions in areas such as design verification, control of the shape and location of elements, replication of tooling and restoration of those products for which the original technical documentation is missing or incomplete.

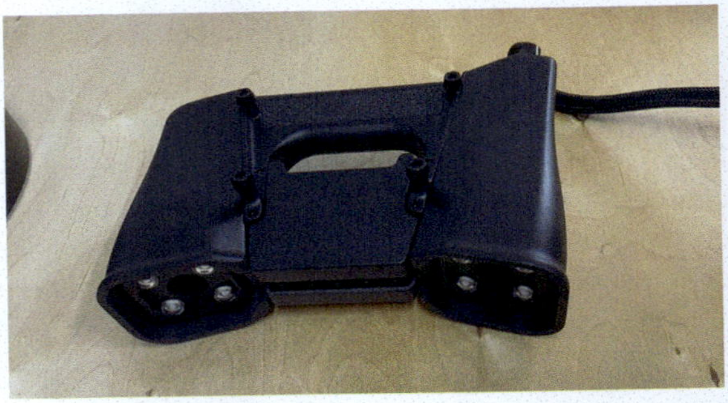

Fig. 1 Scanform L5 Handheld 3D scanner

The main function of the 3D scanner is to simplify the manual collection of high-resolution three-dimensional data on complex, often non-planar surfaces, eliminating the need for stationary installations or turntables. Due to this characteristic, the scanner is well suited for work in the field and when scanning large-sized products. Hand-held scanners differ from coordinate measuring machines in their increased spatial flexibility. This is due to their light weight (700–1000 g), ergonomic controls and the ability to scan objects, ranging from small mechanical parts with characteristic dimensions of about 100 mm to large structures up to 4 m long. Unlike coordinate measuring machines that operate under controlled conditions and with limited measurement ranges, handheld 3D scanners have spatial flexibility. This versatility is additionally provided by a wide single-pass scanning area measuring 250 by 350 mm, allowing the operator to obtain extensive surface data in a single motion path. This is very important when scanning indestructible or fragile objects that cannot be moved during the inspection [3].

In fact, 3D scanners serve as a link between the digital counterparts of objects and the physical objects themselves, being metrological-class tools for creating STL or OBJ grid representations suitable for import into CAD/CAM systems or models for finite element engineering analysis (FEA). In addition, the mobility of such devices allows them to be used locally, for example, on production lines, on construction sites or inside the machines themselves, which expands the possibilities of its use outside the laboratory or metrological laboratory in real conditions where flexibility and speed are crucial.

2 Materials and Methods

The operation of manual 3D scanners in most cases is based on the use of a visible laser line, which is usually in the wavelength range of 635–660 nm. This laser line is projected onto the surface of the object and is detected by a calibrated CMOS sensor. In this study, a scanner of the Scanform L5 model was used.

When the scanner is manually moved over the surface of an object, the displacement of the laser profile in the sensor's field of view is continuously measured. Using triangulation geometry (using a fixed baseline and known optical parameters), the system calculates a dense point cloud that reconstructs the object's topography with submillimeter resolution. In practice, this makes it possible to obtain surface characteristics with a resolution of approximately 0.14 mm per point. This is sufficient for most reverse engineering tasks in automotive and tractor construction.

The scanner device is housed in a rigid composite housing, the geometric dimensions of which are 173 mm long, 94 mm high and 298 mm wide. This ensures a stable design for the internal optical components. With a total weight of about 950 g, the scanner remains light enough to provide long-term manual operation without significant operator fatigue, which is a fundamental parameter in the field and when scanning multiple objects [4].

Internally, the scanner is configured in accordance with the classical laser triangulation architecture, which includes three main subsystems: a laser projection module, a photodetector camera (sensor unit) and an internal structural frame providing fixed angular positioning between these two subsystems. The laser unit emits a coherent beam in the red visible spectrum with a nominal wavelength of about 635 nm. This beam is expanded into a laser line using a cylindrical lens or a diffractive optical element. This line is then projected onto the surface of the scanned object. The orientation and angle of divergence of the laser plane are carefully adjusted to ensure consistent triangulation conditions and the same depth of field throughout the scanning area.

The CMOS image sensor, which forms the core of the detection system, is equipped with a high-resolution matrix capable of detecting small movements of the laser line in accordance with the relief of the object's surface. The angular displacement between the laser projection axis and the optical detection axis is typically between 20° and 30°, which is the optimal configuration for depth sensitivity and minimal occlusion. The baseline, defined as the physical distance between the laser emitter and the camera sensor, is calibrated and remains constant during operation. This baseline forms a fixed part of the triangulation triangle used in the depth calculation algorithm. Each captured frame is subsequently processed in real time to obtain a dense cross-sectional profile. This profile is then summarized along the trajectory of the scanner's manual movement to restore a complete three-dimensional surface.

In addition, the optical components are mechanically isolated from possible vibration interference by internal damping structures, which ensures image stability during movement. Pre-calibrated lens correction models stored in the device's built-in software compensate for optical distortion, and the temperature-stabilized circuit design reduces the effect of thermal expansion. The scanner's built-in data processing system synchronizes the laser pulse and image acquisition with millisecond accuracy, ensuring spatial consistency at all measured points [5, 6].

The Scanform L5 handheld 3D scanner is based on the optical laser triangulation method, which is widely known in the metrological instrumentation industry due to its ability to provide high-precision spatial measurements in a non-contact manner. In this particular case, laser triangulation allows the scanner to determine the depth (Z-coordinate) of each scanned point. This is achieved by geometrically resolving the displacement of the projected laser line on the object's surface relative to the stationary image sensor.

The essence of the triangulation process is to measure the angular displacement of the laser line image on the sensor matrix. Each pixel on the sensor corresponds to a certain angle of incidence and is determined by a known geometric configuration (i.e., a fixed baseline B, a known projection angle α and a calibrated focal length f), it becomes possible to calculate the depth coordinate of any Z illuminated surface point using trigonometric inversion. Mathematically, the principle can be expressed by the following simplified equation:

$$Z = \frac{B \cdot f}{x' - x_0}, \tag{1}$$

where:

- x'—position of the laser line on the sensor;
- x_0—the reference (central) pixel coordinate corresponding to the zero depth.

This process is repeated for each scan line captured by the device as the operator manually moves the scanner over the surface of the object. The scanner captures hundreds to thousands of cross-sectional profiles per second, depending on the sensor's frame rate, which is typically 30–60 frames per second (fps). Each profile contains from 600 to 2000 data points, which provides a theoretical data collection rate of up to 200,000 points per second. However, this speed depends on the reflectivity of the surface, the ambient lighting conditions, and the scanning speed.

The angular configuration of the scanner is optimized to ensure a balance between depth resolution and measurement stability, usually using an observation angle θ of 20° to 30° relative to the laser plane. This geometry provides sufficient inequality for accurate triangulation, while avoiding excessive angular distortion or overlap.

The data collection and calibration process of a handheld 3D scanner is a procedure that ensures high accuracy, geometric accuracy, and reproducibility of digital surface models under various operating conditions. The process is divided into two interrelated subsystems: dynamic data collection during scanning and pre-calibration of the system. Calibration is performed either at the factory or in the field using custom calibration procedures.

To ensure accurate alignment of successive profiles, the scanner implements one of two tracking mechanisms. The main method is to register frames based on geometry using iterative search algorithms for the closest points to align overlapping objects between neighboring frames with pixel accuracy. Conversely, in cases where objects exhibit insufficient geometric contrast or surfaces are reflective, the system can use photogrammetric markers. These markers, which are fixing elements located on the scanned surface, facilitate spatial correlation [7]. Marker-based tracking has been shown to increase the accuracy of recording, reaching an accuracy of $\pm$ 0.05 mm on adjacent frames. This method proved to be particularly effective in scenarios involving scanning larger and more featureless geometric objects.

Recalibration may be required in the field under certain conditions, such as temperature shifts exceeding $\pm$ 5 °C, mechanical shocks, or after 100–150 h of operation. The scanner is equipped with a built-in calibration program that helps the user in a structured process that includes scanning a known calibration object under specified conditions. Deviations from the expected profiles are processed to adjust the internal projection matrices and compensate for any minor mechanical deviations that may have occurred during processing or transportation.

An important component of the data collection process is the system's ability to manage environmental noise. The Scanform L5 camera is equipped with active filtering to eliminate side reflections, dynamic exposure control to adjust ambient light changes (in the range from 300 to 2000 lx), and additional frame exclusion criteria to remove low-confidence data segments. Together, these mechanisms provide a high signal-to-noise ratio of the resulting point cloud, usually exceeding

45 dB, which is necessary for subsequent grid reconstruction and dimensional analysis.

The 3D scanner has a balanced set of operating parameters that ensure high accuracy, versatility and ease of use in various operating conditions, especially in the field of reverse engineering. Its spatial resolution is maintained at approximately 0.14 mm, which makes it possible to capture the fine surface details needed to reconstruct geometrically complex mechanical components. The depth resolution reaches ± 0.05 mm under controlled conditions, while the overall surface accuracy does not exceed ± 0.1 mm for objects longer than 1 m when tracking the marker and following proper calibration protocols [8]. The scanner's field of view per frame is 250 by 350 mm, which provides a scanning area of 87,500 mm^2 and provides effective scanning at a working distance of 250–450 mm. The depth of field is approximately 150 mm, which allows for sequential triangulation on curved or uneven surfaces. With a frame capture rate of 30–60 frames per second and an average of 1800 dots per frame, the scanner provides a throughput of 54,000 to 108,000 dots per second. Reliable operation is ensured in ambient conditions from + 10 °C to + 35 °C and relative humidity up to 80%, without condensation, with reliable control of ambient light from 300 to 2000 lx. Calibration remains stable for at least 100 h of operation, and field recalibration can be performed in 5–8 min while maintaining repeatability within ± 0.03 mm [9, 10]. Data export to standard formats such as STL, OBJ, and PLY is supported, with resulting mesh files ranging in size from 5 to 150 MB, depending on the scan complexity and resolution. The scanner is connected via USB 3.0 and consumes no more than 12 W of electricity during continuous operation.

The technical specifications of the Scanform L5 scanner are shown in Table 1.

3 Results

Before starting the scanning process, the surface of the object must be carefully treated to ensure optimal laser reflectivity and geometric traceability. The surface preparation process includes:

- Surface cleaning: The scanned object should be free of dust, oil, condensation, or dirt that can disperse or absorb the directed laser beam. The cleaning process is usually performed using isopropyl alcohol or antistatic wipes to prevent the appearance of residual streaks.
- If necessary, it is recommended to apply a matte finish to achieve optimal results. A thin layer of scanning aerosol is applied to parts with glossy, mirrored or translucent surfaces (for example, chrome, glass, ABS plastic). This coating usually contains titanium dioxide or other compounds with microparticles that serve to reduce specular reflection and improve the optical properties of the surface. It is extremely important to maintain the thickness of the layer within the optimal range of 10–15 microns, since excessive coating can lead to deformation of small geometric elements.

Table 1 Technical characteristics of the Scanform L5 scanner

Parameter	Meaning
Spatial resolution	~ 0.14 mm
Depth resolution (Z-axis)	± 0.05 mm
Global surface accuracy	± 0.1 mm (for objects ≥ 1 m)
Field of view in one frame	250 mm × 350 mm
Scan area per frame	87,500 mm^2
Working distance	250–450 mm
Depth of field	~ 150 mm
Frame capture rate	30–60 frames/s
Data collection rate	54,000–108,000 dots/s
Scanner dimensions	173 mm × 94 mm × 298 mm
Scanner weight	950 g
Operating temperature range	+ 10 °C to + 35 °C
Permissible relative humidity	≤ 80% (non-condensing)
Resistance to external light	300–2000 lx
Calibration stability	≥ 100 h of operation
Recalibration time	5–8 min
Measurement error	± 0.03 mm
Supported output formats	STL, OBJ, PLY
Typical file size with the model	5–150 MB
Interface and connection	USB 3.0
Power consumption	≤ 12W

- The placement of markers, if applicable, is an important aspect of the methodology. In cases where an object lacks a natural geometry for element-based alignment, retroreflective markers are evenly distributed over its surface. The placement density should correspond to one marker per 80–150 cm^2 of surface area, with at least three markers visible in each frame for triangulation alignment. It is extremely important that the markers are applied without skewing and without overlapping edges, joints or elements of high curvature.

Fixing the part: In cases where scanning from different angles is required or the object needs to be reoriented, it is important to use a stable mount or a turntable to ensure accurate positioning. It is extremely important that the object's coordinate system matches the scanner's reference frame.

A "revolver" was used as the scanning object, which houses nozzles for spraying pesticides from an agricultural sprayer (see Fig. 2).

After preparing the part, the scanner must be connected to the main computer, and then run the software. The system automatically checks the calibration status,

Fig. 2 The sprayer part prepared for scanning

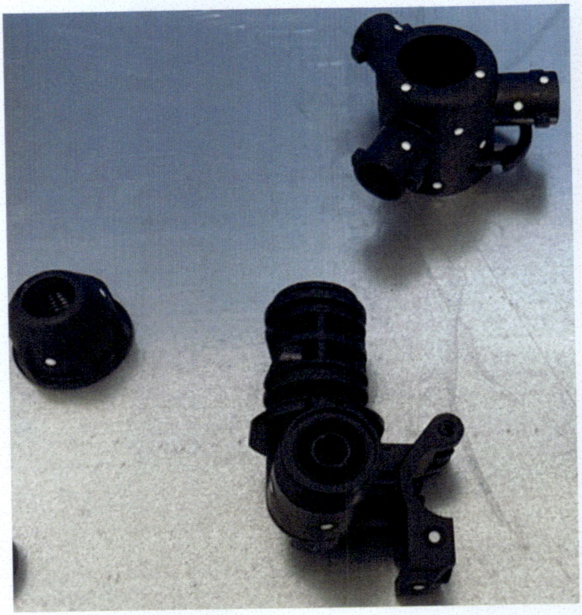

confirming the internal geometric parameters (camera characteristics, baseline, orientation of the laser plane). If the deviation thresholds are exceeded (due to mechanical action or temperature), an interactive recalibration procedure using a certified instrument is started, with a typical execution time of 5–8 min.

During operation, the scanner is located at a distance of approximately 370 mm from the surface of the object, which corresponds to the optimal operating range of triangulation from 250 to 450 mm (see Fig. 3). The built-in multi-color LED indicators help the operator determine the optimal range, as indicated by the green indicator, excessive proximity, as indicated by the red indicator, and the distance of the scanner to the target, as indicated by the blue indicator. The scanner should be positioned at an angle of 20–30° to the surface normal to ensure optimal visibility of the strip and prevent blockage.

The device projects five finely focused red laser lines onto the surface of the target. Due to the fact that they coincide with the topographic features of the local landscape, these lines are detected by a CMOS sensor. Each captured frame allows you to obtain a point profile of a cross-section with a frame rate of 30–60 frames per second and a density of 1200–2000 dots per frame, depending on the reflectivity of the surface and the speed of movement of the scanner.

The operator runs the scanner over the surface in slow, uniform movements at a speed of 100–150 mm/s, ensuring an overlap of at least 30% from frame to frame (see Fig. 4). The system registers each captured profile using:

- Iterative nearest point alignment algorithm for surface geometry-based alignment;

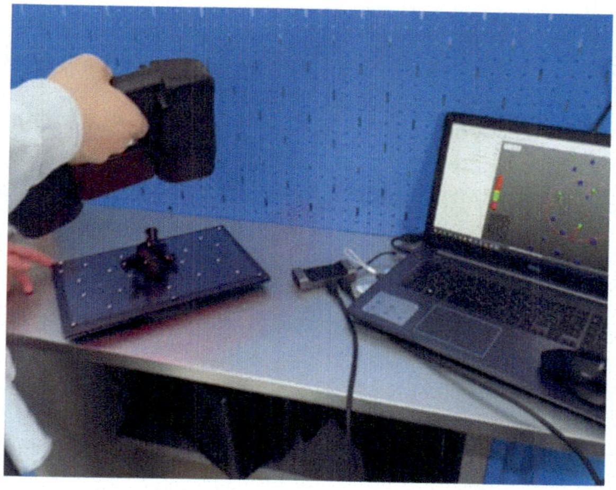

Fig. 3 The sprayer part prepared for scanning

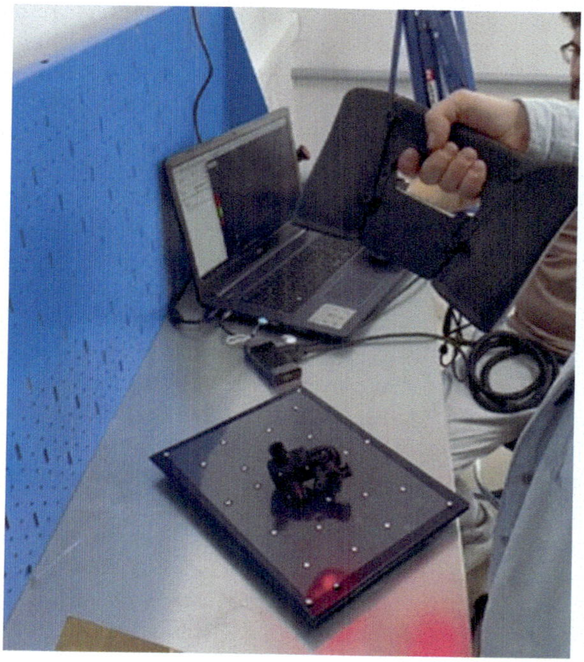

Fig. 4 The process of scanning an object

- In the presence of markers, fiduciary data-based registration ensures alignment accuracy up to ± 0.05 mm.

The software simultaneously visualizes a changing point cloud in 3D, providing information about coverage, overlaps, and error vectors. Areas with insufficient data can be re-scanned before completion.

Thus, the quality of the 3D scanning result depends not only on the optical equipment, but also on the preparation of the object. Before scanning, it is necessary to check the cleanliness, optical characteristics of the surface, stability and visibility of the image. When configured correctly, this scanner allows you to obtain accurate 3D models in shape and size, suitable for subsequent modeling tasks.

After completing the scanning process using a portable device, the received data goes through a comprehensive and technically structured post-processing cycle. The cycle is designed to transform a raw point cloud into a geometrically accurate and topologically reliable digital model. This step is necessary to eliminate inconsistencies, align multiple scan paths, eliminate surface gaps, and create waterproof meshes suitable for use in engineering design, analysis, or digital manufacturing.

The initial result of the scanner is a dense, unstructured point cloud containing three-dimensional spatial coordinates, often reaching hundreds of thousands or even millions of data points, depending on the size of the object and the resolution of the scan. This cloud may contain various defects, including misaligned scan frames, noisy data points due to reflective or dark surfaces, incomplete areas due to overlap, or redundant data due to overlapping passages. Therefore, at the initial stage of subsequent processing, it is necessary to fully register all individual scanning stages in a single spatial coordinate system. In cases where photogrammetric markers were used during the scanning process, the software automatically aligns all segments with a high degree of accuracy. This alignment is achieved by using known spatial relationships between markers, which ensures frame-to-frame consistency with an error of $\pm$ 0.05 mm. In the absence of markers, the software uses geometric algorithms, such as iterative registration of nearby points, to align adjacent frames by using overlapping surface elements. The effectiveness of this step depends on sufficient profile overlap, usually exceeding 30%, to ensure numerical stability.

After the equalization is completed, the data is processed by noise reduction and statistical outlier removal. This operation allows you to filter out points that differ significantly from the local geometric structure, which is often the result of diffraction at the edges, high reflectivity, or external interference, by analyzing the distance between each point and its nearest neighbors. Data integrity is further enhanced by the use of smoothing filters such as sliding least squares or Laplace smoothing, which preserve the geometry of the macro scale while reducing high-frequency surface noise. At this stage, the software calculates the surface normals for each point, which are necessary for the correct orientation of the triangle when building the grid.

One of the most serious problems with post-processing is the elimination of partial gaps and inaccessible geometries, which lead to the formation of holes on the surface. If these gaps are within a reasonable range (usually less than 5 mm), they are filled automatically using curvature-based interpolation algorithms. Larger holes may require manual intervention using grid editing software, in which the operator defines the boundaries and applies NURBS or spline-based corrections. Once the point cloud is completely cleared and filled, it is transformed into a polygonal surface grid using triangulation. The Scanform L5 system usually uses Poisson surface reconstruction methods or Delaunay methods to create a topologically consistent grid. The resulting grid consists of uniformly oriented triangles with an average edge length of 0.1–0.3 mm, which ensures compatibility with both high-precision control tasks and subsequent design solutions.

After creating the grid, optimization procedures are performed to reduce the number of polygons, where possible, without compromising the accuracy of the surface image. This includes thinning, correcting edges unrelated to the manifold, and smoothing the normal vector. In the context of reverse engineering, the mesh can be imported into CAD software, where it serves as the basis for constructing parametric models by fitting the surface and defining geometric constraints. In these workflows, the dimensional deviation between the reconstructed mesh and the final CAD model is quantified through best fit analysis, with tolerances typically maintained within ± 0.1 mm for mechanical applications.

Geomagic Design X is a sophisticated industrial software platform that has been carefully designed to transform high–resolution 3D scanning data into precise parametric CAD models. This software is widely used in reverse engineering processes, especially when converting scanned point clouds or polygonal grids into editable solid-state models based on the simplest elements. These models are designed to provide full compatibility with major CAD systems, including SolidWorks, Siemens NX, PTC Creo, and Autodesk Inventor.

After importing a grid or point cloud into Geomagic Design X, the software performs a series of data preparation steps. These steps include thinning the grid, removing noise, smoothing, and segmenting areas. These tools allow users to reduce the amount of unnecessary data while maintaining geometric accuracy, usually with tolerances in the range of ± 0.05–0.1 mm. The software provides both automatic and manual surface extraction tools for the identification and reconstruction of primitive geometries (planes, cylinders, spheres) and NURBS surfaces of arbitrary shape. A distinctive feature of this software is its ability to transfer the reconstructed geometry to its own CAD environments in real time, accompanied by a detailed analysis of the parameters. This capability ensures the smooth integration of reverse engineering components into existing design workflows, facilitating collaboration and improving the efficiency of project teams.

In Design X, the model can be refined using advanced tools such as deviation analysis, which compares the CAD reconstruction with the original scan data, highlighting local deviations using color maps (see. Figure 5). This functionality ensures geometric alignment and is essential for quality assurance in high-precision design areas. In addition, the software facilitates hybrid modeling by integrating scanned surfaces with built-in CAD features such as holes, protrusions, and curves. This functionality is especially useful when reverse engineering components.

4 Discussion

The present study confirms the effectiveness of using a portable 3D scanner in combination with an automated design system for high-precision reverse engineering of agricultural machinery components. The applied methodology makes it possible to

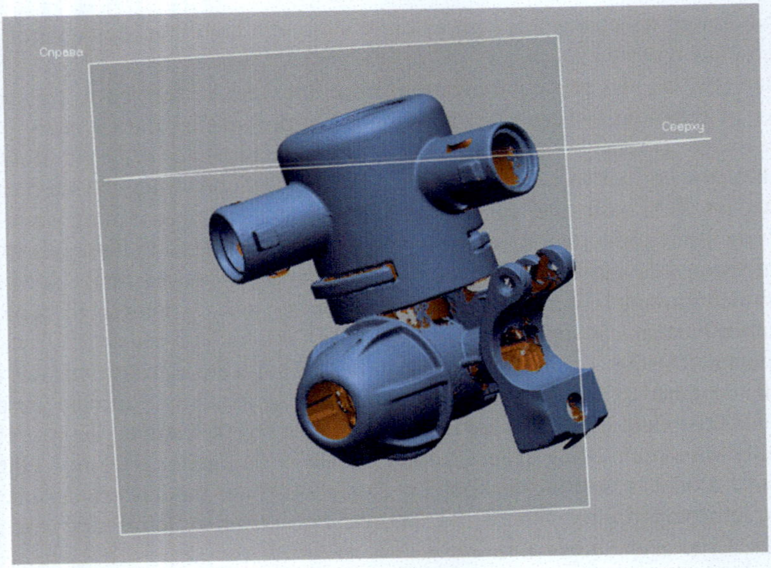

Fig. 5 The process of scanning an object

accurately reconstruct complex geometric shapes with a spatial resolution of up to 0.14 mm and depth accuracy within ± 0.05 mm. The system is highly mobile, repeatable, and reliable in the field, making it suitable for scanning large and irregular parts. The subsequent processing and reconstruction in CAD of these components ensures their full integration into the work processes of engineers, facilitating the digital restoration and re-manufacture of important components. This approach is scalable and adaptable for broader applications in mechanical engineering, especially in industries requiring import substitution and reduced downtime.

5 Conclusion

Based on the results of the study, a justification was formed for the practical applicability and metrological applicability of a combined workflow combining structured 3D laser scanning with subsequent digital surface reconstruction and parametric modeling. The developed approach provides geometrically accurate reproduction of complex components of agricultural machinery, many of which are otherwise unavailable due to supply chain failures, obsolescence, or lack of technical documentation.

Empirical data has shown that the Scanform L5 scanner, operating on the principle of laser triangulation, generates a high-density point cloud with an average spatial resolution of approximately 0.14 mm and depth accuracy within ± 0.05 mm. This level of accuracy makes it possible to capture important morphological features on non-planar, wear-prone surfaces and surfaces made of various materials, such as seed dispenser shafts, sprayer brackets and gear housings. The scanning area of the system, combined with its portability and real-time visual feedback, allows large components to be scanned directly in the field or in the workshop, eliminating the need for rigid mounts or optical targets.

The implementation of a calibration protocol capable of maintaining submillimeter stability for operating intervals exceeding 100 h, and marker-based registration procedures ensure consistent and reproducible data collection even under changing environmental conditions. In addition, the post-processing process, which includes noise filtering, triangulated mesh reconstruction, topology reconstruction, and CAD model generation, has proven effective in creating solid-state digital counterparts suitable for direct integration into CAE and CAM environments. The use of Geomagic Design X made it easier to extract primitive geometry and perform surface modeling of arbitrary shape, accompanied by an analysis of deviations that were maintained within a tolerance of $\pm$ 0.1 mm on reconstructed surfaces.

The implemented reverse engineering cycle thus bridges the technological gap between outdated physical components and modern digital production processes. This facilitates the transition from damaged or undocumented mechanical parts to editable 3D geometry optimized for re-fabrication or structural redesign. The workflow can be further extended to various branches of engineering, in addition to agriculture, including the automotive and aerospace industries, where dimensional accuracy and geometric shape restoration are of key importance.

In conclusion, it is worth noting that the integration of the laser scanning system into the reverse engineering process provides a reliable, scalable and cost-effective solution for the restoration and redesign of important mechanical elements. This makes it possible to solve current production problems related to import substitution, thereby increasing the stability of tractor fleets in the face of technological and economic constraints.

Acknowledgements This article was prepared as part of research work, implemented at the expense of the federal budget (funding source - the Ministry of Education and Science of the Russian Federation) scientific topic code FZNW-2024-0026.

References

1. Badakova, V.A.V., Taldykin, D.S., Pletnev, M.G.: Influence of annealing on impact strength of machine parts manufactured by 3D printing from polylactide. Traktory i selkhozmashiny **91**(6), 819–826 (2024). https://doi.org/10.17816/0321-4443-636691
2. Nefelov, I.S., Filatov, V.V., Yudin, D.A.: Influence of 3D scanning data processing methods on the quality of final solid 3D models. Standarty i kachestvo **11**, 76–77 (2024). https://doi.org/10.35400/0038-9692-2024-11-257-24
3. Bykov, V.V., Golubev, M.I.: Prospects of using 3D scanning technologies in quality control of machine parts. In: Innovative Technologies in the Agro-industrial Complex of the Region:

Achievements, Problems, Development Prospects, Proceedings of the National Scientific and Practical Conference, Tver, Russia, 9–11 Feb 2021. Tver State Agricultural Academy, Tver, pp. 320–323 (2021)

4. Bendik, N.V., Kuporzhanov, D.E.: Development of a 3D scanning system for agricultural machine parts. In: Applied Aspects of Mathematics and Natural Sciences in Education, Engineering and Economics, Proceedings of the International Scientific and Practical Conference. Dedicated to the 90th Anniversary of the Department of Mathematics, Irkutsk, Russia, 23–24 May 2024, pp. 206–211. Irkutsk State Agrarian University named after A.A. Ezhevsky, Molodezhny (2024).
5. Sablin, A.I., Farkhutdinov, I.M., Arslanov, V.V.: Application of 3D scanning as a method of incoming control of components and units for urban and agricultural machinery. In: Young Scientists for Innovative Development of the Agro-Industrial Complex, Proceedings of the XI National and Scientific and Practical Conference of Young Scientists, Ufa, Russia, 4 Dec 2018, Part I, Bashkir State Agrarian University, Ufa, pp. 249–253 (2018)
6. Suprun, T.I., Saneev, M.I.: Features of modern technologies used in the design of machine parts at the production stage. In: Problems and Prospects for the Implementation of Interdisciplinary Research, Proceedings of the International Scientific and Practical Conference, Voronezh, Russia, 15 May 2023, LLC "Agency for International Research", Sterlitamak, pp. 155–160 (2023)
7. Markov, V.A., Karasev, Yu.A.: reverse engineering in forestry machinery manufacturing. In: Advanced Technologies of Reverse Engineering and Rapid Prototyping, Proceedings of All-Russian Scientific and Practical Conference, Sevastopol, Russia, 13–15 May 2024, Sevastopol State University, Sevastopol, pp. 86–88 (2024).
8. Bernikov, M.A., Solovyov, M.V.: Methodology of reverse engineering for housing parts of rotary machines: development and testing. Cifra. Mashinostroenie **2**(7) (2025). https://doi.org/10.60797/ENGIN.2025.7.2
9. Yunusbaev, N.M.: 3D Scanning in the technology of repairing tractor and vehicle components and assemblies. Sel'skokhozyaystvennye mashiny i tekhnologii **13**(1), 4–8 (2019). https://doi.org/10.22314/2073-7599-2018-13-1-4-8
10. Kasimov, R.M., Tuzhilin, S.P., Arumugam, G.: Opportunities for using 3D scanners in the agro-industrial complex. Electrotechnol. Electr. Equip. Agric. **68**(4), 103–111 (2021). https://doi.org/10.22314/2658-4859-2021-68-4-103-111

Modeling Climate Impact on Agricultural Systems Using Machine Learning Methods

Sharshenbek Zhusupkeldiev, Chyngyz Sabitov, Rahat Sagyndykova, Damira Asanbekova, and Gulzat Ismailova

Abstract In this research, various models based on machine learning algorithms and neural networks were created to predict air quality. Various regression models were constructed to predict the air quality index (AQI). For the regression task, a comparative analysis of modern gradient boosting algorithms for AQI forecasting, including XGBoost, LightGBM, and CatBoost, was conducted. A comparative analysis of the baseline and optimized models was conducted. Regression curves were plotted, and their performance was analyzed. A forecasting comparison of the performance of neural networks with the Adam and RMSProp optimizers for predicting air quality was conducted. A multivariate regression model was constructed and optimized. To improve model quality, neural network hyperparameters were optimized. A neural network was constructed, and the results were compared with the Adam and RMSProp optimizers. Their accuracy and model error were assessed, and error matrices and ROC curves were constructed, showing air quality. Using the Random Forest model, we predicted the concentration of AQI (or other particulate matter) based on data on $PM_{2,5}$, PM_{10}, CH_4, CO_2, NO_2, SO_2, pollutants, meteorological parameters (temperature, humidity, wind speed, etc.), and temporal variables. Nonlinear and hidden relationships between the variables and AQI were identified using the traditional Pearson correlation matrix when the data are normally distributed, and Spearman and Kendall rank correlation matrices when the data are not normally distributed. Temporal patterns of AQI autocorrelation and cross-correlation with meteorological parameters were determined.

S. Zhusupkeldiev (✉) · G. Ismailova
Kyrgyz National University named after J. Balasagyn, Bishkek, Kyrgyzstan
e-mail: sharshen58@mail.ru

C. Sabitov
Kyrgyz State University named after I. Arabaev, Bishkek, Kyrgyzstan

R. Sagyndykova
Kyrgyz National Agrarian University named after K.I. Scriabin, Bishkek, Kyrgyzstan

D. Asanbekova
Kyrgyz State Medical Academy named after I. Akhunbaev, Bishkek, Kyrgyzstan

D. Muhamediyeva et al. (eds.), *Engineering and Digital Technologies in Agriculture*, Smart Agriculture 20, https://doi.org/10.1007/978-981-92-0481-6_5

Keywords Air quality · Machine learning methods · Neural networks · Neural network optimizers · Linear and nonlinear modeling · Forecasting · Model accuracy · Comparative analysis

1 Introduction

In response to climate change, actions aimed at sustainable air pollution management. The objectives of this research are to improve monitoring and reduce emissions of ground-level air pollutants. The primary objective is to develop cooperation in achieving sustainable development goals in agricultural regions. The objective here is to predict the environmental impact on the quality of agricultural products, which is crucial for food production. Predicting air pollutant concentrations during periods of high pollution is crucial, a key requirement for environmental management and its impact on agricultural systems. Accurate forecasting models can positively impact public health protection efforts by facilitating timely decision-making to prevent exposure to high concentrations of air pollutants. Outdoor air pollution is a serious public health concern due to its harmful impact on populations and the quality of produce grown by farmers and agricultural producers. Air pollution is a complex mixture of particles, chemicals, and biological materials. Air pollutants mainly consist of particulate matter ($PM_{2.5}$, PM_{10} and ozone (O_3), volatile organic compounds, nitrogen dioxide (NO_2), carbon monoxide (CO), sulfur dioxide (SO_2) [1]. At elevated concentrations, pollutants O_3, $PM_{2.5}$ (fine particulate matter) and NO_2, even cause mortality with long-term exposure to the human body [2, 3]. Long-term exposure to $PM_{2.5}$ and PM_{10} pollutants has been associated with chronic cerebrovascular, pulmonary, cardiac, pneumonia, lung cancer, and respiratory diseases [1, 4, 5]. CO and SO_2 emissions are mainly generated by industrial facilities during solid fuel combustion and incomplete combustion in vehicle engines, gas equipment, wood, and fossil fuel combustion [6]. Scientific studies have found significant health effects from CO exposure, including myocardial ischemia, cardiac arrhythmia, and even suffocation or death at high concentrations [7–9]. Furthermore, studies have also linked CO exposure in pregnant women to fetal thrombosis, stillbirth, preterm birth, and, in some cases, congenital heart disease [10–13].

Traditional air quality monitoring systems or information systems based on Internet of Things (IoT) technologies provide real-time air quality indicators, which is an important requirement for air pollution management. Therefore, these systems can support data-driven decisions to protect people from adverse health effects when a certain air pollutant or air quality index (AQI) exceeds permissible limits [14]. Accordingly, the development of air pollutant concentration prediction models allows us to anticipate the pollution trend for the next few hours or days, helping in environmental management decisions to reduce risks and prevent human exposure to highly concentrated environmental pollution. The next step in air quality systems is to develop air pollution prediction systems for hourly or daily air pollutant concentrations with a high level of accuracy and reliability, which will be a great advantage for governments or agencies, supporting appropriate decision-making to protect human health during environmental emergencies [15].

Thus, air pollutant concentration levels (e.g., $PM_{2.5}$ or PM_{10}) can be represented as time series data, that is, a sequence of observations at specified time intervals (hourly, 8-h average, or daily, depending on the standard governing the air pollutant and the analysis to be conducted), within which linear or nonlinear components and behavior within the time series data can be identified, whether they are trending, cyclical, seasonal, or irregular. Time series representing air pollutant concentration values have random characteristics, and the data may be linear, nonlinear, or not normally distributed. Therefore, identifying patterns in these data is a challenging task. Such behavior of air pollutant data can be associated with natural sources characterizing the study area and anthropogenic sources of air pollution caused by human activities. To build air quality models based on deep learning, it is important to utilize various deep learning architectures. This technology has the ability to identify complex nonlinear relationships, which has proven effective in improving the discriminative performance of models. Neural network modeling techniques using complex time series data can improve forecasting performance for complex categories of greenhouse gas impacts on air quality. Recurrent neural networks (RNNs) are a key component of these architectures and are known for their highly effective forecasting models with time series data. RNNs utilize a sequence of input data with cyclical connections between units, where neurons are connected to each other in the same hidden layer, repeatedly applying a learning function to the hidden states. In particular, long short-term memory (LSTM) neural networks, a type of RNN, have demonstrated remarkable capabilities in learning long-term dependencies, a crucial aspect of sequential data. They also effectively address issues of nonlinearity, periodicity, seasonality, and serial dependency between data sequences, as seen in air pollutant concentration levels. LSTM models are commonly applied to sequence and time series forecasting tasks in various fields, as well as to predict the process of small particle air pollution and the impact of meteorological factors.

2 Materials and Methods

Main problems solved by machine learning methods. Predicting pollutant concentrations.

Linear regression. Objective function. Predicting a continuous variable y based on a linear combination of features x. A linear regression model can be represented as

$$y = w^T x + b \tag{1}$$

where w is the weight and b is the bias.

The optimization of the model is based on gradient descent, the model of which is based on the optimization problem through the loss function (MSE) in the form

$$\mathcal{L}(w, b) = \frac{1}{N}\sum_{i=1}^{N}\left(y_i - w^T x_i - b\right)^2 \rightarrow \min \tag{2}$$

where the unknown parameters of the model are updated according to recurrence relations

$$b^n = b^{n-1} - \eta\nabla_b\mathcal{L} \tag{3}$$

$$w^n = w^{n-1} - \eta\nabla_w\mathcal{L}\left(w^{n-1}, b\right) \tag{4}$$

where η is the learning rate. Algorithms that produce high prediction rates include gradient boosting (XGBoost, LightGBM). They are typically used for complex data to achieve high accuracy results. The minimization problem with a loss function and regularization for this method is defined as

$$\mathcal{L} = \sum_{i=1}^{N} L(y_i, \hat{y}_i) + \sum_{m=1}^{M}\Omega(f_m), \quad \Omega(\mathrm{f}) = \gamma\mathrm{T} + \frac{1}{2}\lambda||w||^2, \tag{5}$$

where $L(y_i, \hat{y}_i)$ is the quality functional. We now describe some key aspects of the mathematical foundation of neural networks and deep learning. One of the key methods for the initial construction of a fully connected neural network (MLP) is a fully connected feedforward network; the formula is used to transfer data between neurons. The Kendall and Spearman rank correlation coefficients are commonly used to analyze statistical relationships. Let $(x_1, y_1), \ldots, (x_n, y_n)$ be a sample of observations of a two-dimensional random vector (X, Y) with distribution $F_{XY}(x, y)$. The Kendall rank correlation coefficient is a rank statistic defined by the following expression:

$$\tau_{XY} = \frac{N^+ - N^-}{\frac{1}{2}n(n-1)} \tag{6}$$

where N+ is the number of pairs of observations (x_i, y_i), (x_j, y_j), $i > j$, for which the condition $(x_i - x_j)(y_i - y_j) > 0$, is satisfied, N– is the number of pairs of observations (x_i, y_i), (x_j, y_j), $i > j$, for which the condition $(x_i - x_j)(y_i - y_j) < 0$ is satisfied. In other words, N+ is the number of observed pairs that have the same tendency to change in both features: either an increase in the value of one increases the value of the other, or a decrease in the value of one decreases the value of the other. N– is the number of observed pairs with opposite tendencies to change. Let us sort the observation results in ascending order of the values of feature X. Then the sample of ranks of feature X will be a sequence of natural numbers 1, 2, ..., n (if all observations $x_1, \ldots, x_n$ are different). We denote the corresponding sample of feature Y ranks as $s_1, \ldots, s_n$. In practice, the following formula is used to calculate the sample value of the Kendall rank correlation coefficient:

$$\tilde{\tau}_{XY} = \frac{4Q}{n(n-1)} - 1 \tag{7}$$

where

$$Q = \sum_{i=1}^{n-1} Q_{i,}\ Q_i = \sum_{j=i+1}^{n} [s_j - s_i] \tag{8}$$

The number of ranks in the sample $s_{i+1}, \ldots, s_n$, greater than s_i.

Using Formula (7) gives the correct result only for the case when there are no repeating elements in the samples $x_1, \ldots, x_n$ and $y_1, \ldots, y_n$. However, with a small number of repeating elements, this error can be neglected in practice. The properties and interpretation of the Kendall rank correlation coefficient are similar to the properties and interpretation of the Spearman rank correlation coefficient. Thus, for a functional monotonically increasing dependence between random variables X and Y, $\tau_{XY} = 1$, and for a monotonically decreasing dependence: $\tau_{XY} = -1$. For independent random variables X and Y, the mathematical expectation $M[\tau_{XY}] = 0$. Sample values of the Spearman correlation coefficient, as a rule, are higher (in absolute value) than the sample values of the Kendall correlation coefficient. This effect is associated with a greater sensitivity of the first coefficient to the discrepancy in the trends of changes in the values of the features. To test the significance of the Spearman rank correlation coefficient, we formulate the main hypothesis:

$$H_0 : \tau_{XY} = 0 \tag{9}$$

The criterion statistic used is the statistic:

$$Z = \tilde{\tau}_{XY}\sqrt{\frac{9n(n+1)}{2(2n+5)}} \tag{10}$$

which, given the truth of H_0 and a large sample size for $n > 30$, is approximated by the standardized normal distribution:

$$f_Z(z|H_0) \sim N(0, 1) \tag{11}$$

In the deep learning section of machine learning, the general problem statement is formalized as finding an optimal mapping (function) from the feature space to the response space, using the mapping $f{:}X \rightarrow Y$, where $X \subseteq R^d$ is the feature space, Y is the response space (targets), and $f \in F$ is the desired function from the class of admissible functions. The Adam optimizer is important for optimizing neural networks and has numerous practical applications. Let's consider its mathematical justification: being an adaptive optimization, when updating its parameters using recurrent formulas, it uses the Momentum and RMSprop optimization methods, which take into account a combination of the idea of gradient inertia and an adaptive learning rate for each parameter. Adam individually calculates adaptive learning rates for each parameter, while estimating the mean gradient and the uncentered variance of gradients, which correspond to the Momentum and RMSprop methods. We will describe the operating principle of the Adam algorithm. For each parameter θ, it calculates the first moment, the average gradient $m_0 = 0$, and calculates the second moment, the non-centered variance $v_0 = 0$. A counter t is introduced, and for each t, the gradient of the objective loss minimizing function is calculated using the formula $g_t = \nabla_\theta J(\theta_{t-1})$:

$$J(\boldsymbol{\theta}) = \frac{1}{m}\sum_{i=1}^{m} L\left(f\left(x^{(i)}, \boldsymbol{\theta}\right), y^{(i)}\right) \tag{12}$$

by $\boldsymbol{\theta} = (\theta_1, \theta_2, \ldots, \theta_n)$, θ is an unknown vector, $y^{(i)} = f(x^{(i)})$ is the target variable we want to predict, $\hat{\boldsymbol{\theta}} = \left(\hat{\theta}_1, \hat{\theta}_2, \ldots, \hat{\theta}_n\right)$ is the minimum point of the functional (3.1), and $f_i(x^{(i)}, \theta)$ is the model we want to build. Next, the recurrence formula of the first moment is updated:

$$m_t = \beta_1 m_{t-1} + (1 - \beta_1)\, g_t, \tag{13}$$

where $\beta_1 \in [0,1]$ is called the damping coefficient, the initial value is usually taken to be equal to 0.9. Next, the second moment is updated and the moment offset is corrected. The formula for the non-centered moment, i.e., the second moment, is determined as follows.

$$\upsilon_t = \beta_1 \upsilon_{t-1} + (1 - \beta_1)\, {g_t}^2 \tag{14}$$

where $\beta_2 \in [0,1]$ is called the damping coefficient for the second moment, the initial value is usually taken to be 0.999, ${g_t}^2$- means element-wise squaring. The moment shift update is carried out according to the formulas:

$$\hat{m}_t = m_t / \left(1 - \beta_1^t\right), \hat{\upsilon}_t = \upsilon_t / \left(1 - \beta_2^t\right) \tag{15}$$

We update the recurrence formula:

$$\theta_t = \theta_{t-1} - \alpha \frac{\hat{m}_t}{\sqrt{\hat{\upsilon}_t} + \varepsilon} \tag{16}$$

where α is the learning rate, $\varepsilon \approx 10^{-8}$ is a small constant for the stability of the calculations of the elements of the recurrence relations. The first moment, as an exponentially decaying moving average of the gradients, similar to the method of moments, helps accelerate convergence in stable directions and suppresses fluctuations in the loss function. The second moment, an exponentially decaying moving average of the squared gradients, allows adapting the model's learning rate for each parameter. It decreases the learning step for parameters with large gradients. It increases the step for parameters with small gradients, which helps avoid local minima. Bias correction is necessary since the initial values $m_0 = v_0 = 0$ lead to biased estimates at the beginning of training. A major advantage and adaptability of Adam is the automatic tuning of the model's learning rate for each parameter. It stores different update scales for different parameters. To save memory, it stores only the moments m and v for each parameter (O(n)), which saves memory. Maintains gradient invariance when updating and scaling. The Adam optimizer is robust to noisy gradients and sparse gradient matrices.

3 Results and Discussion

Using neural network modeling and machine learning algorithms, we constructed model learning curves with loss visualizations on the training and validation sets. Common error matrices were constructed for various classification models, including binary classification. ROC-AUC analysis was conducted for the constructed models. Key regression metrics (MAE, MSE, RMSE, R^2) were used to evaluate model accuracy, along with error statistics (mean, standard deviation, skewness, and kurtosis). Regression (CO level forecasting). The Adam and RMSProp optimizers showed similar results in terms of MSE and R^2. The Adam optimizer often demonstrates smoother training and lower error variability during validation. RMSprop can converge faster at the beginning of training. For air quality classification, Adam demonstrated slightly better results in model accuracy and AUC. RMSprop can produce comparable results, but requires more careful tuning of the model training speed. Both models perform well in class discrimination (AUC > 0.85). Training stability was observed with the Adam optimizer, which demonstrated more stable training with smaller loss function fluctuations. RMSprop may be more sensitive to the choice of learning rate. Key aspects of the Adam and RMSprop optimizers are discussed, showing that while they are reliable choices, additional parameter tuning may be required to achieve comparable results. L2 regularization was implemented for all dense layers, with a dropout probability of 0.3 and batch normalization to stabilize training (Fig. 1).

To predict positively defined classes, error matrices and ROC-AUC curves were constructed to determine the quality of the model using the random forest machine learning algorithm (see Fig. 2).

For the constructed optimized models using Adam and RMSProp for the predicted classes, the positive definite classes are shown in Fig. 3.

In the analysis of the classification problem in Listing 1, the random forest algorithm yielded slight improvements in prediction results compared to the neural network.

Listing 1. Comparison of the classification of the full AQI air quality report using a neural network and the random forest algorithm.

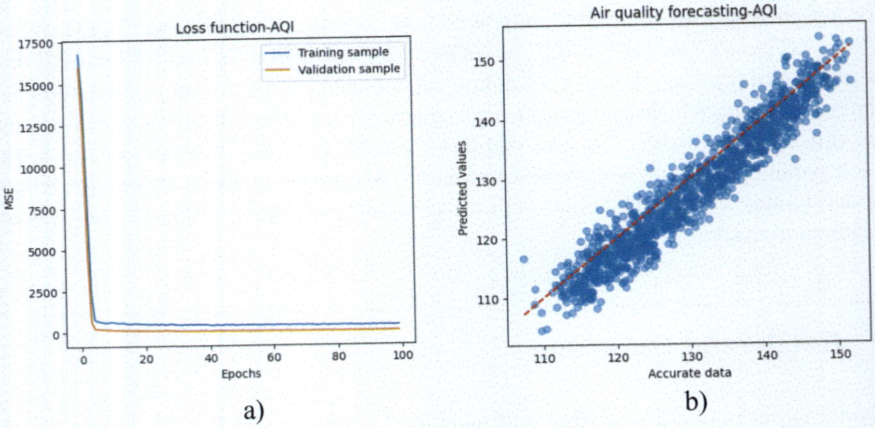

Fig. 1 Neural network training errors **a** and multivariate regression model **b**

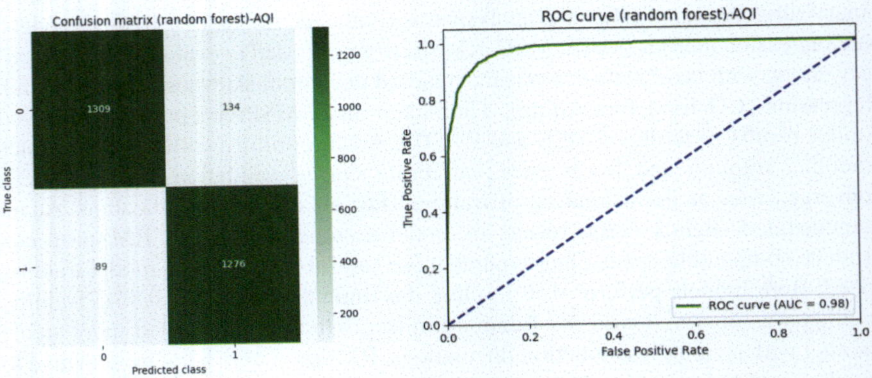

Fig. 2 Error matrix and ROC curves for random forest implementation

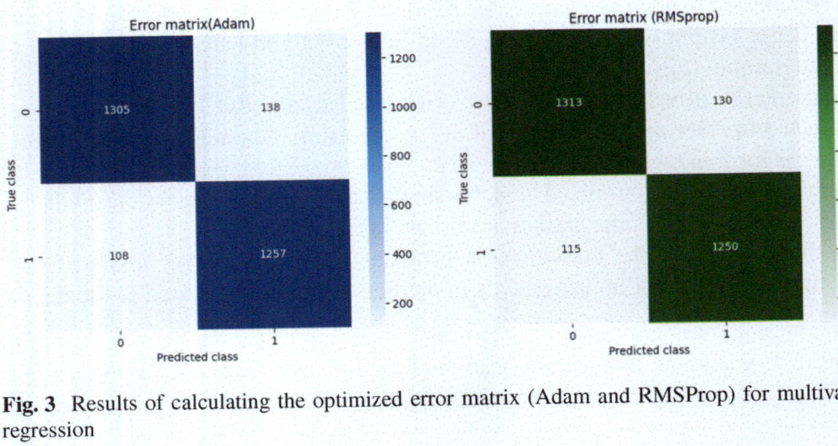

Fig. 3 Results of calculating the optimized error matrix (Adam and RMSProp) for multivariate regression

```
Neural network (classification)-AQI:
              precision    recall  f1-score   support

           0       0.93      0.90      0.92      1443
           1       0.90      0.93      0.92      1365

    accuracy                           0.92      2808
   macro avg       0.92      0.92      0.92      2808
weighted avg       0.92      0.92      0.92      2808
```

```
Random forest (classification):
              precision    recall  f1-score   support

           0       0.94      0.91      0.92      1443
           1       0.90      0.93      0.92      1365

    accuracy                           0.92      2808
   macro avg       0.92      0.92      0.92      2808
weighted avg       0.92      0.92      0.92      2808
```

The results of constructing regression models and assessing model quality using R^2, constructed using the XGBoost, LightGBM, and CatBoost algorithms, shown in Fig. 4, demonstrated high performance, with a nearly 100% predictive value for model quality. All constructed models had R^2 values approaching 1, indicating an ideal model.

Below are the results of MSE, RMSE, and R^2 evaluations for machine learning models: linear regression, gradient boosting, random forest, decision tree, and support vector machine. Figure 5 shows that RMSE error estimation is superior to MSE model estimation. R^2 estimation shows relatively good results.

When constructing a regression model using a neural network and the random forest method, the results of correctly classified data on the ROC curve tend toward the upper corner of 1, indicating a perfect model. However, misclassified data tend

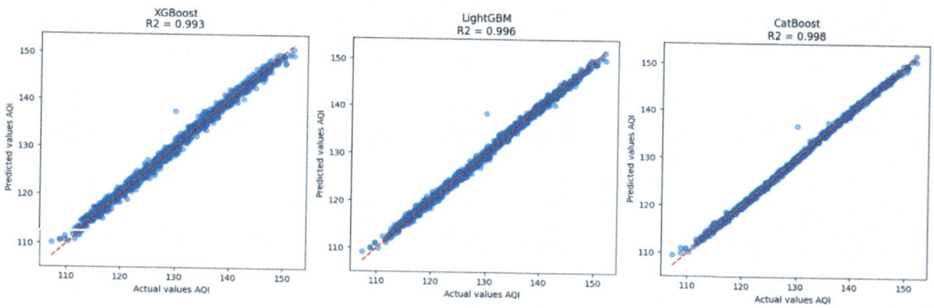

Fig. 4 Performance of XGBoost, LightGBM and CatBoost regression models

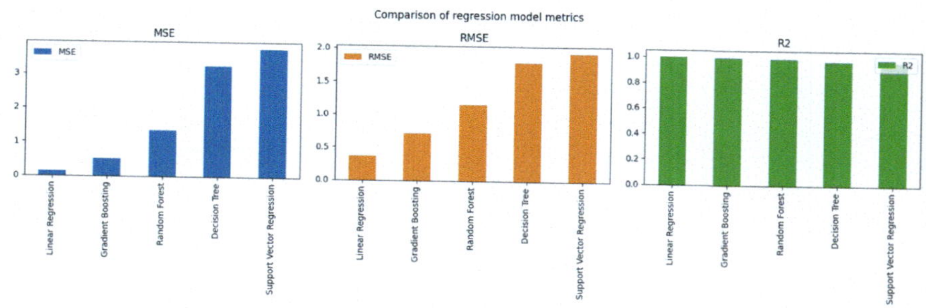

Fig. 5 Metric features for assessing the quality of machine learning models in MSE, RMSE, and R^2 metrics

toward 0. Importantly, when optimizing models, the results show similar accuracy (Fig. 6).

Results were obtained for a model constructed using a neural network with the Adam and RMSProp optimizers for the AQI regression model and conventional gradient boosting with optimized gradient boosting and updated optimized hyperparameters after tuning. In the first case, it is clear that the neural network model has large deviations from the data of the optimized regression model using Adam and RMSProp (see Fig. 5a).

Figure 7 shows the results of neural modeling of the regression model, the obtained curves of error and training accuracy for optimized models based on the Adam and RMSProp optimizers on training and test data sets.

A full forecasting report based on the optimized model using the Adam and RMSProp optimizers was generated, demonstrating 98% accuracy in predicting air quality. Figure 8b shows that, for linear regression, the meteorological parameter precipitation has the greatest impact on air quality, increasing air quality. Other

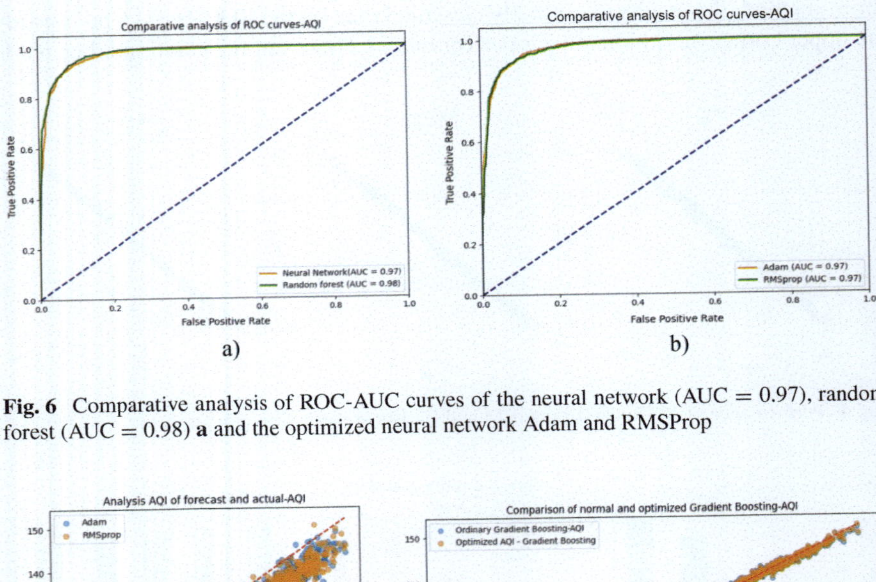

Fig. 6 Comparative analysis of ROC-AUC curves of the neural network (AUC = 0.97), random forest (AUC = 0.98) **a** and the optimized neural network Adam and RMSProp

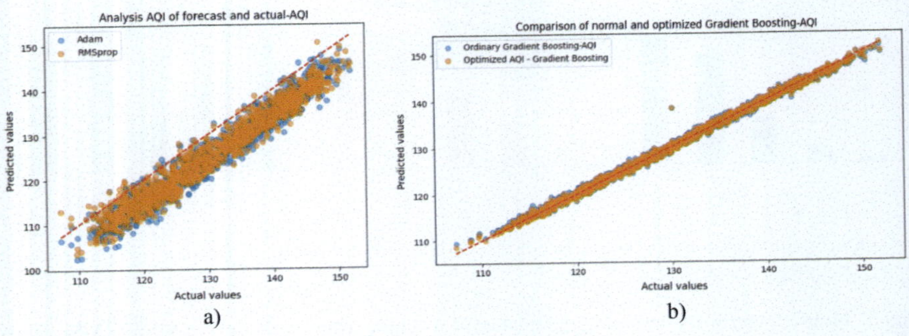

Fig. 7 Optimized regression model and regression model built using gradient boosting with hyperparameter tuning

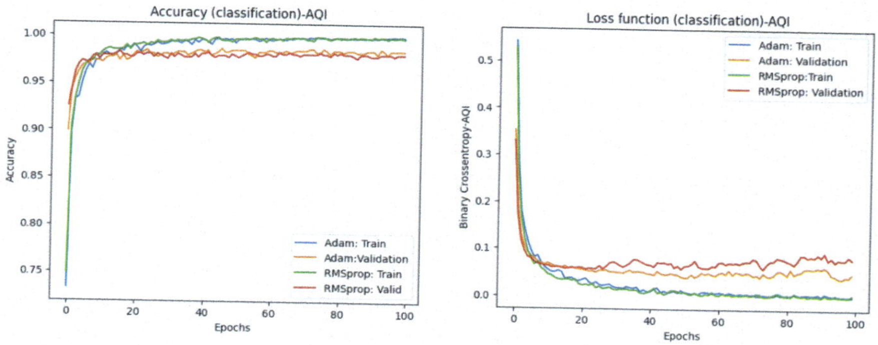

Fig. 8 Results of learning curves of optimized models

```
Classification results (Adam):-AQI
              precision    recall  f1-score      support
0              0.984266  0.985989  0.985127   571.000000
1              0.985428  0.983636  0.984531   550.000000
accuracy       0.984835  0.984835  0.984835     0.984835
macro avg      0.984847  0.984813  0.984829  1121.000000
weighted avg   0.984836  0.984835  0.984835  1121.000000

Classification results (RMSprop):-AQI
              precision    recall  f1-score      support
0              0.985965  0.984238  0.985101   571.000000
1              0.983666  0.985455  0.984559   550.000000
accuracy       0.984835  0.984835  0.984835     0.984835
macro avg      0.984815  0.984846  0.984830  1121.000000
weighted avg   0.984837  0.984835  0.984835  1121.000000
```

a)

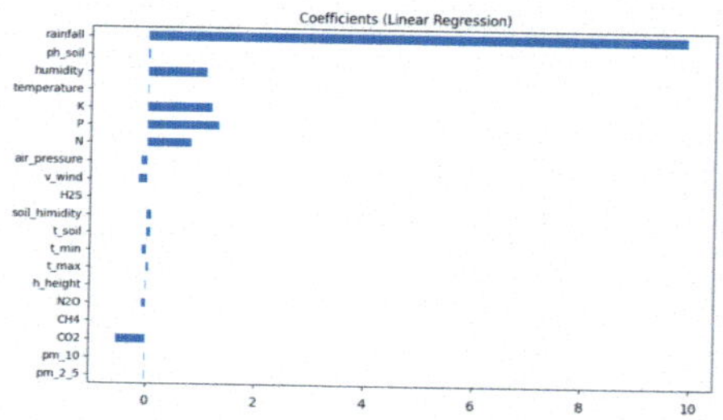

b)

Fig. 9 Full report of the importance of features of the optimized model Fig. 9 **a** and the influence of meteorological parameters and other parameters on the coefficients of the linear regression model Fig. 9 **b**

parameters, such as air humidity and the agronomic soil parameters nitrogen, phosphorus, and potassium, moderately increase AQI, while CO_2, N_2O decrease AQI. Wind speed and soil moisture negatively impact AQI, thus increasing air quality.

Figure 10a–e show the results of implementing the machine learning regression models: linear regression, decision tree, and advanced random forest, gradient boosting, and support vector machine algorithms. According to the calculation results, the coefficients of determination of the air quality regression models showed $R^2 = 0.999$ for linear regression, indicating that the ideal model was constructed; decision tree $R^2 = 0.969$; random forest $R^2 = 0.987$; gradient boosting $R^2 = 0.995$; and support vector machine showed the result $R^2 = 0.965$. The performance of all models. The regression models almost approached 1, indicating the ideality of all constructed models. Almost all models showed a denser grouping of data around the main diagonal, which means that the constructed model differs little from the scattered data. The mutual dependence between the signs and AQI is greatest with precipitation, soil temperature and moderate with nutrients K-potassium, P-phosphorus, there are dependencies on the variables of altitude, air pressure, CO_2 and soil acidity.

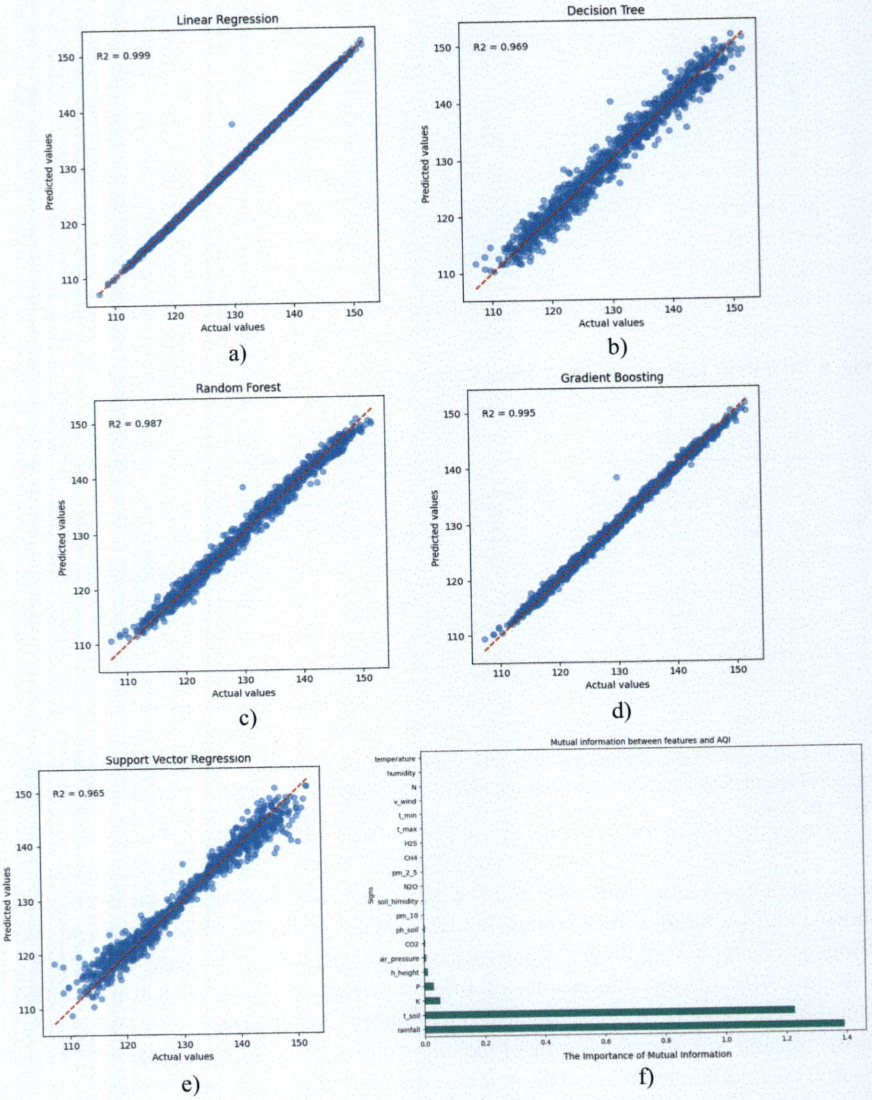

Fig. 10 Machine learning regression models Fig. 10**a–e** and the interdependence between feature functions with AQI

To identify hidden dependencies between features and the AQI, autoencoder neural network models were constructed. The training error results for this model are shown in Fig. 11. It is demonstrated that, for large epochs, the model learns well and tends to achieve high accuracy with minimal model error. To identify hidden correlations between features and the AQI, latent features latent_1—latent_6 were identified (Fig. 11b).

The constructed model (see Fig. 12a) shows that the AQI value has a temporal dependence, which can be regulated by influencing values from previous moments. Existing temporal patterns and statistical relationships between features are associated with the autocorrelation values with the AQI and can persist for a certain time (see Fig. 12a, b) shows the dependence of the AQI with five main features (see Fig. 12b). Figure 13 shows the importance of features determined by the XGBoost machine learning algorithm.

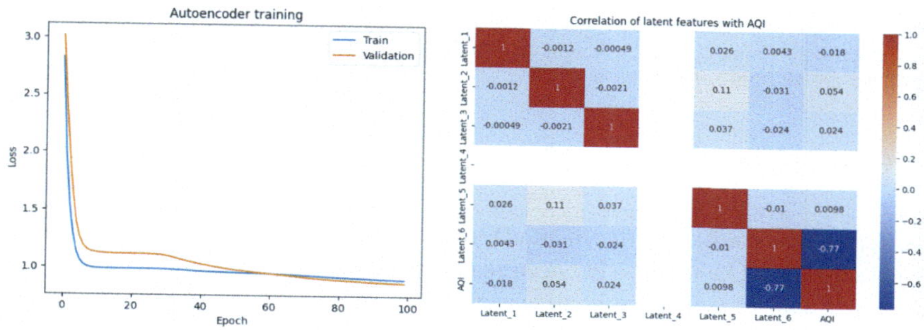

a) b)

Fig. 11 Error of the autoencoder model for identifying hidden dependencies between features on training and validation data Fig. 11 **a**. Correlation between hidden features and with air quality AQI Fig. 11 **b**

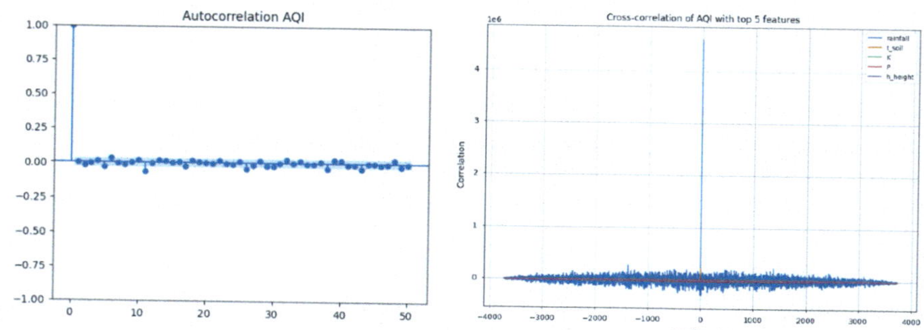

a) b)

Fig. 12 The constructed AQI autocorrelation model (Fig. 12**a**) and the importance of the top 5 main features

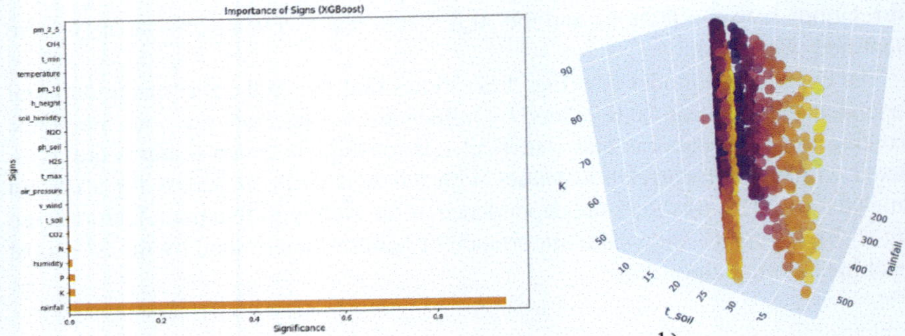

a) b)

Fig. 13 Feature importance shows that precipitation and correlation relationships between top 5 features-rainfall, t_soil, K, P and h_height with AQI and 3D visualization of top 3 features-rainfall, t_soil and potash fertilizer-K with AQI

4 Conclusion

The simplest and most effective method for creating a model for predicting air quality, especially if the relationships between features are close to linear, is a linear regression algorithm. For complex nonlinear data, ensemble machine learning methods, feedforward neural networks, or a dense-layer neural network architecture such as DenseNet are better. Comparative analyses of two models using random forest and neural networks for regression tasks yielded the following results. For predicting CO levels, both models showed comparable results. Random forest may have a slight advantage in terms of R^2. For the classification task (good/poor air quality), the predominantly random forest showed slightly better results in terms of AUC. The neural network may require additional hyperparameter tuning. The constructed ROC curves showed that both models separate classes quite well, but the random forest performs slightly better. Using feature importance analysis in the random forest, we showed which air features are most significant for AQI prediction. The CatBoost algorithm showed the best results both in the basic version with $R^2 = 0.921$ and after optimization with $R^2 = 0.928$. The LightGBM algorithm demonstrated comparable performance with $R^2 = 0.919$ after optimization. The advanced XGBoost algorithm is slightly inferior to its competitors with $R^2 = 0.913$ after optimization. Feature importance is determined in Fig. 8f) based on XGBoost. The greatest efficiency increase in quality after optimization was provided by XGBoost optimization up to $+0.008\ R^2$. Another algorithm, CatBoost, showed the smallest increase of $+ 0.007\ R^2$, due to its initially high performance. The LightGBM algorithm improved its performance by $+0.007\ R^2$. The regression curves for all constructed models demonstrated good agreement between predicted values and actual accurate values. In the CatBoost regression model, the scatter showed a denser clustering of points around the diagonal. In the region of high CO values (right side of the graphs), all models slightly underpredicted. The following feature importance results were demonstrated. The following features had the main influence on CO levels: nitrogen oxides (NOx), air temperature fluctuations, and humidity. For this task, the CatBoost and LightGBM algorithms are the optimal choices, which may be preferable when working with large data sets. The XGBoost machine learning algorithm is used if compatibility with legacy systems is important. Figure 12 the cross-correlation of AQI with 5 main characteristics was established. Lags, i.e., interrelations by lag or shift of features among themselves, were checked. Basic linear and nonlinear main results on the interrelationship of key variables and their correlating properties were obtained. Linear dependencies were established between the pm_2_5, pm_10 and CO_2 features with AQI, which showed an ROC curve with the determination coefficient $r = 0.82$, $r = 0.79$ and $r = 0.75$, respectively. Important hidden interrelations of mutual information were obtained between t_max (MI = 0.62), v_wind (MI = 0.58), humidity (MI = 0.55). Unexpected dependencies were discovered, in the form of a nonlinear relationship between ph_soil and AQI with MI = 0.48 and $r = 0.12$. Air pressure (air_pressure) showed a weak linear relationship with the AQI, but a significant nonlinear relationship with the AQI. Current AQI values were found to depend on previous values at different points in time, and there were temporal patterns, statistical relationships between features, and the relationship between autocorrelation values and the AQI persisted for up to 24 h. The maximum cross-correlation with meteorological parameters was observed at a lag of 3–5 h. 3D visualizations of the three best characteristics—rainfall, soil temperature, and potassium fertilizer application—were established with the AQI.

References

1. Delavar, M.A., Jahani, M.A., Sepidarkish, M., Alidoost, S., Mehdinezhad, H., Farhadi, Z.: Relationship between fine particulate matter (PM2.5) concentration and risk of hospitalization due to chronic obstructive pulmonary disease: a systematic review and meta-analysis. BMC Public Health. **23**, 2229 (2023)
2. Anenberg, S.C., Mohegh, A., Goldberg, D.L., Kerr, G.H., Brauer, M., Burkart, K., Hystad, P., Larkin, A., Wozniak, S., Lamsal, L.: Long-term trends in urban NO_2 concentrations and associated pediatric asthma incidence: estimates from global datasets. Lancet Planet. Health. **6**, e49–e58 (2022)
3. Andreão, W.L., de Almeida, T., Albuquerque, T.: Avoidable mortality by implementing more restrictive fine particles standards in Brazil: an estimation using satellite surface data. Environ. Res. **192**, 110288 (2021)
4. Domingo, J.L., Rovira, J.: Effects of air pollutants on the transmission and severity of respiratory viral infections. Environ. Res. **187**, 109650 (2020)
5. Gutman, L., Pauly, V., Orleans, V., Piga, D., Channac, Y., Armengaud, A., Boyer, L., Papazian, L.: Long-term exposure to ambient air pollution is associated with an increased incidence and mortality of acute respiratory distress syndrome in a large French region. Environ. Res. **212**, 113383 (2022)
6. Liu, C., Yin, P., Chen, R., Meng, X., Wang, L., Niu, Y., Lin, Z., Liu, Y., Liu, J., Qi, J.: Ambient carbon monoxide and cardiovascular mortality: a nationwide time-series analysis in 272 cities in China. Lancet Planet. Health. **2**, e12–e18 (2018)

7. Liu, Y., You, J., Dong, J., Wang, J., Bao, H.: Ambient carbon monoxide and relative risk of daily hospital outpatient visits for respiratory diseases in Lanzhou, China. Int. J. Biometeorol. **67**, 1913–1925 (2023)
8. Taheri, M., Nouri, F., Ziaddini, M., Rabiei, K., Pourmoghaddas, A., Islam, S.M.S., Sarrafzadegan, N.: Ambient carbon monoxide and cardiovascular-related hospital admissions: a time-series analysis. Front. Physiol. **14**, 1–9 (2023)
9. Goldsborough, E., Gopal, M., McEvoy, J.W., Blumenthal, R.S., Jacobsen, A.P.: Pollution and cardiovascular health: a contemporary review of morbidity and implications for planetary health. Am. Heart J. Plus Cardiol. Res. Pract. **25**, 100231 (2023)
10. Chillrud, S.N., Ae-Ngibise, K.A., Gould, C.F., Owusu-Agyei, S., Mujtaba, M., Manu, G., Burkart, K., Kinney, P.L., Quinn, A., Jack, D.W.: The effect of clean cooking interventions on mother and child personal exposure to air pollution: results from the Ghana randomized air pollution and health study (GRAPHS). J. Expo. Sci. Environ. Epidemiol. **31**, 683–698 (2021)
11. Kaali, S., Jack, D.W., Mujtaba, M.N., Chillrud, S.N., Ae-Ngibise, K.A., Kinney, P.L., Boamah Kaali, E., Gennings, C., Colicino, E., Osei, M.: Identifying sensitive windows of prenatal household air pollution on birth weight and infant pneumonia risk to inform future interventions. Environ. Int. **178**, 108062 (2023)
12. Alexander, D.A., Northcross, A., Karrison, T., Morhasson-Bello, O., Wilson, N., Atalabi, O.M., Dutta, A., Adu, D., Ibigbami, T., Olamijulo, J.: Pregnancy outcomes and ethanol cook stove intervention: a randomized-controlled trial in Ibadan, Nigeria. Environ. Int. **111**, 152–163 (2018)
13. Wylie, B.J., Kishashu, Y., Matechi, E., Zhou, Z., Coull, B., Abioye, A.I., Dionisio, K.L., Mugusi, F., Premji, Z., Fawzi, W.: Maternal exposure to carbon monoxide and fine particulate matter during pregnancy in an urban Tanzanian cohort. Indoor Air. **27**, 136–146 (2017)
14. Méndez, M., Merayo, M.G., Núñez, M.: Machine learning algorithms to forecast air quality: a survey. Artif. Intell. Rev. **56**, 10031–10066 (2023)
15. Bekkar, A., Hssina, B., Douzi, S., Douzi, K.: Air-pollution prediction in smart city, deep learning approach. J. Big Data. **8**, 161 (2021)

Application of Artificial Intelligence Methods in Precision Agriculture for Arid Regions

Narly Babanazarov, Ilmyrat Ilyasov, Serdar Dowletov, Guwanch Akyyev, and Gurbanbibi Orazbayeva

Abstract In this study, the application of artificial intelligence methods in precision agriculture for arid regions is analyzed under conditions of chronic water scarcity and high climatic variability. The role of multi-source data, including satellite indicators, soil moisture measurements, meteorological observations, and irrigation parameters, is investigated to improve the reliability of environmental monitoring and forecasting. Early drought signals, vegetation stress patterns, and rapid soil moisture depletion are identified using a hybrid analytical framework based on deep learning models. The integration of remote sensing indices with ground-based sensor data is examined to enhance real-time estimation of soil moisture conditions and crop yield in arid agroecosystems. The effect of multi-modal data fusion on the predictive accuracy of crop performance, irrigation demand, and drought-related anomalies is determined in comparison with conventional regression-based approaches. A methodological framework for an artificial intelligence-enhanced agricultural digital twin is established to support continuous data assimilation, model recalibration, and adaptive forecasting under changing climatic conditions. A hybrid predictive workflow combining temporal neural networks, agro-climatic modeling, and anomaly detection modules is developed to strengthen early warning systems for drought and heat stress in water-limited farming systems. The practical use of artificial intelligence for optimizing irrigation scheduling, resource allocation, and climate-resilient crop management is proposed for operational decision-support tools in agriculture. An analytical environment that is scalable, interpretable, and robust enough to support sustainable agriculture in arid and semi-arid regions is presented.

Keywords Precision agriculture · Artificial intelligence · Soil moisture forecasting · Remote sensing analytics · Vegetation stress detection · Arid regions · Hybrid deep learning models · Agronomic decision support

---

N. Babanazarov (✉) · I. Ilyasov · S. Dowletov
Turkmen State Institute of Economics and Management, Ashgabat, Turkmenistan

G. Akyyev
Institute of Telecommunications and informatics of Turkmenistan, Ashgabat, Turkmenistan

G. Orazbayeva
Dovletmammet Azadi Turkmen National Institute of World Languages, Ashgabat, Turkmenistan

D. Muhamediyeva et al. (eds.), *Engineering and Digital Technologies in Agriculture*, Smart Agriculture 20, https://doi.org/10.1007/978-981-92-0481-6_6

1 Introduction

1.1 Relevance and Research Problem

The fast transformation of agricultural systems due to digitalisation, climate instability and resource constraint has radically transformed traditional methods of crop management, yield forecasting and environmental monitoring. Modern agricultural landscapes, especially in arid places, generate huge amounts of different data from remote sensing platforms, soil moisture sensors, unmanned aerial vehicles, climate monitoring stations and satellite-based vegetation indices. Because standard agronomic technologies cannot capture non-linear environmental interactions, dynamic soil-plant-atmosphere relationships, and the multi-factor character of crop development under severe water limits, these data streams are too complicated and irregular to analyze [1, 2]. In order to achieve sustainable productivity under these circumstances, sophisticated analytical models, multi-source data fusion technologies, and artificial intelligence techniques that can generate accurate, real-time, adaptive insights must be integrated.

The notion of precision agriculture, which focuses site-specific management and the optimizing of resource usage based on data, has emerged as one of the most important methods to solving these difficulties. However, typical precision agricultural approaches often become ineffective when used to desert ecosystems defined by significant temperature variability, chronic water deficits, soil salinity and rapid climatic variations. This is because of their insufficient capacity to integrate real-time sensor feedback into predictive frameworks, poor representation of spatial variability, and inadequate modeling of microclimatic dynamics [3]. Consequently, conventional techniques struggle to give correct recommendations for irrigation scheduling, drought stress detection, and crop resilience assessment in the severe and extremely variable circumstances typical of desert landscapes.

Recent advances in artificial intelligence, including machine learning, deep neural networks, hybrid spatiotemporal models and adaptive forecasting algorithms, have paved the way for the development of intelligent precision agriculture systems that can learn complex environmental patterns and respond dynamically to rapidly changing field conditions. Artificial Intelligence-driven techniques can find hidden connections between meteorological variables, soil moisture profiles, plant physiological responses, and remote sensing indicators when compared to rule- or regression-based models. This provides greatly better accuracy when estimating crop stress, evapotranspiration, yield formation and water demand [4–6]. Adopting scalable cloud infrastructures and edge-computing sensor networks further boosts the ability of such systems to function continuously, recalibrate model parameters in real time and support long-term agricultural planning.

Despite tremendous progress, numerous crucial obstacles remain unresolved. Firstly, agricultural activities in arid places are highly impacted by abrupt climatic phenomena such as heat waves, dust storms, dry winds and intermittent droughts, which are tough to incorporate into static or weakly adaptable Artificial Intelligence models [7]. Secondly, combining heterogeneous data sources ranging from satellite imaging to in-field Internet of Things measurements requires standardized structures and robust pre-processing pipelines, which are typically lacking in existing precision agricultural systems. Thirdly, the poor interpretability of machine learning predictions limits their practical adoption by decision-makers and agronomists who need clear, empirically supported explanations of model outputs [8].

These restrictions underline the urgent demand for a unified methodological framework that can combine data-driven Artificial Intelligence models, domain-specific agronomic expertise and scalable information infrastructures. While maintaining interpretability and resilience, this system must facilitate ongoing data collection, multi-layer analytical processing, real-time prediction, and context-aware decision-making. An Artificial Intelligence-enhanced precision agricultural system should offer insights on crop stress evolution, soil moisture dynamics, ideal irrigation techniques, and resilience-boosting mechanisms, especially in arid regions.

The need to create an integrated, Artificial Intelligence-based precision agriculture architecture that is especially suited to the operational, technological, and environmental limitations of dry ecosystems determines the study's relevance. The proposed strategy intends to bridge the gap between theoretical developments in agricultural data science and the practical deployment of intelligent decision support systems. It delivers a scalable, transparent and highly accurate approach for enhancing crop productivity, optimizing resource allocation and strengthening sustainability in the face of adverse climatic conditions.

1.2 Literature Review

The rapid increase of artificial intelligence applications in agriculture has led to a major rise in academic interest in precision farming technologies, particularly in places confronting chronic water shortages and adverse weather conditions. Modern research continuously reveal that Artificial Intelligence-enhanced analytical frameworks outperform traditional agronomic models by offering more accurate assessments of crop stress, soil moisture dynamics, evapotranspiration rates and yield changes in arid environments. According to Zhang et al. [9], integrating machine learning with remote sensing data enables real-time monitoring of field heterogeneity and delivers improved adaptability compared with standard deterministic models. These discoveries have established artificial intelligence as a core feature of next-generation precision agriculture systems.

Recent comparative evaluations have showed that Artificial Intelligence-driven precision agricultural systems offer much higher predictive accuracy in simulating drought, soil-plant interactions and water use optimization. Deep learning techniques, in particular convolutional neural networks and long short-term memory, have proven to be highly effective at modeling temporal dynamics in water-limited situations, interpreting multispectral data, and capturing non-linear crop responses. In comparison to traditional vegetation-index-based approaches, Khan et al. [10] claim that hybrid models that combine satellite imagery with environmental sensor data can enhance crop stress detection by up to 35%. Furthermore, attention-based architectures, such as transformers, have lately been popular for multi-source data fusion in agricultural forecasting tasks, hence broadening the analytical scope of Artificial Intelligence in dry land management even further.

In the meanwhile, a growing body of research is looking at how artificial intelligence might be used to optimize resource allocation, irrigation scheduling, and decision support systems in dry farming systems. Reinforcement learning techniques provide autonomous irrigation control by continually learning the best water-allocation algorithms under changing climatic conditions, as noted by Bhuiyan [11]. Similarly, merging Internet of Things-enabled wireless sensor networks with machine learning gives highly granular measures of soil moisture, canopy temperature and nutrient distribution. Al-Saeed et al. [9] claim that these sensor-Artificial Intelligence frameworks can lower irrigation demand in desert horticulture systems by up to 25% and greatly increase the dependability of water-use efficiency measures. According to these studies, intelligent, networked monitoring infrastructures are becoming more and more important for precision agriculture in water-stressed areas [12].

Artificial Intelligence-driven crop output forecasts and early warning systems for heat and drought stress have also been the subject of several studies. Deep spatio-temporal learning models, including convolutional long short-term memory networks and graph neural networks, have been used to simulate the intricate relationships between soil, climate and crop physiology [13]. For instance, Sharma et al. [14] revealed that hybrid spatio-temporal networks beat classic regression-based yield models in predicting crop performance under high temperature changes. These findings demonstrate how Artificial Intelligence may enhance agricultural resilience by offering incredibly accurate forecasts to assist proactive adaption strategies.

Researchers highlight the persistent constraints related to data scarcity, model interpretability, and environmental unpredictability in dry areas despite substantial progress. Low plant cover, high reflectance noise, and erratic climatic extremes can affect vegetation indices and sensor inputs in desert habitats, making machine learning generalization more difficult. A recent study by Arlanova et al. [15] demonstrates the practical efficiency of artificial intelligence in irrigation control, showing that hybrid machine learning models significantly reduce water consumption while improving crop productivity in arid environments. Their findings highlight how data-driven irrigation scheduling, soil moisture prediction, and vegetation-health monitoring can dramatically enhance resource efficiency compared to rule-based irrigation systems. The results confirm that Artificial Intelligence-enabled smart irrigation is a critical component of precision agriculture, particularly in regions facing chronic water scarcity, providing strong empirical evidence for the integration of predictive analytics into irrigation decision-making frameworks.

Overall, the literature clearly reveals a trend towards data-driven, Artificial Intelligence-enabled precision agriculture frameworks that can address the special issues of dry regions. However, there are still substantial limitations in terms of developing universal methodological standards, integrating multi-modal environmental data and boosting model interpretability for practical field deployment. This study addresses these gaps by creating a conceptual and methodological foundation for applying artificial intelligence to precision agriculture under dry climatic restrictions, delivering a thorough analytical methodology.

1.3 Purpose of Study

The purpose of this study is to look at how agricultural systems in dry areas can be made more accurate, efficient, and sustainable using artificial intelligence approaches. In regions with water scarcity and high ecological vulnerability, special attention is given to finding the best machine learning models, data-driven analytical techniques, and computational frameworks that can capture the complex relationships between soil conditions, climatic variability, irrigation processes, and crop physiological responses. The goal of the project is to provide an integrated methodological framework for applying Artificial Intelligence to enhance agronomic forecasting accuracy, optimize resource allocation, and facilitate prompt decision-making in precision farming workflows.

Additionally, this research aims to establish how multi-modal data sources, such as satellite imagery, soil sensor networks, meteorological sequences and remote sensing indicators, can be combined to create coherent analytical representations that enable the continuous monitoring and predictive assessment of agricultural conditions. Through the exploration of synergies between machine learning algorithms, environmental modelling and geospatial analytics, the study intends to develop a scalable, computationally transparent framework that will enhance the reliability of crop growth predictions, improve irrigation scheduling and boost the resilience of farming systems in arid regions. The study's ultimate goal is to build the methodological instruments required for the real-world application of artificial intelligence in precision agriculture, thereby supporting the long-term growth of agriculture in areas with scarce water supplies.

2 Materials and Methods

This study's methodology is based on a multi-layer strategy that combines machine learning algorithms, remote sensing analytics, agro-climatic modeling, and soil-plant interaction models tailored for dry agricultural settings. In order to capture the nonlinear impacts of drought stress, soil degradation, evapotranspiration variability, and rapid microclimatic changes, precision agriculture in water-scarce locations necessitates constant synchronization of field-level data with predictive algorithms. Therefore, the research framework includes: (1) a conceptual structure of the agricultural digital twin; (2) identification of heterogeneous data sources; (3) mathematical modeling of soil-moisture dynamics and crop responses; (4) integration of artificial intelligence methods for forecasting and decision support.

2.1 Theoretical Framework of the Agricultural Digital Twin

A bidirectional link between real-world farming systems and their virtual analytical counterpart that permits continuous data ingestion, simulation, and Artificial Intelligence-based field condition prediction is known as a "digital twin" in precision agriculture:

- Physical layer—soil, crops, irrigation infrastructure, meteorological conditions;
- Virtual layer—mathematical models, remote-sensing analytics, machine learning predictions;
- Data-exchange layer—real-time sensor streams, satellite pictures, environmental databases.

This framework ensures that agro-ecosystem states are regularly updated in the virtual environment, producing accurate predictions for moisture availability, vegetation stress, biomass development, and irrigation demand.

2.2 Data Sources for the Digital Twin of Arid Agriculture

The study employs a wide range of datasets that thoroughly reflect crop conditions, soil qualities, and environmental dynamics in order to develop a realistic and adaptive analytical model for precision agriculture in arid environments. The primary remote sensing sources include satellite imagery from the Sentinel-2 MultiSpectral Instrument with a spatial resolution of 10–20 m, providing vegetation indices such as the Normalized Difference Vegetation Index, the Normalized Difference Water Index, and the Soil Adjusted Vegetation Index. Additional satellite observations from the Moderate Resolution Imaging Spectroradiometer Terra and Moderate Resolution Imaging Spectroradiometer Aqua platforms give land-surface temperature measurements and evapotranspiration estimations. Landsat-8 Operational Land Imager and Thermal Infrared Sensor data give soil reflectance values and soil moisture indicators necessary for surface-level characterisation.

A second crucial element of the dataset is ground-based measurements. Subsurface water dynamics are captured with great temporal resolution by volumetric soil moisture readings using capacitance-based probes placed at depths between 10 and 60 cm. Local monitoring systems include temperature and humidity sensors to record microclimatic factors that affect crop development. Drip-irrigation infrastructure includes pressure and flow monitoring that enable the identification of irrigation uniformity, system performance, and water-distribution abnormalities.

The dataset is further enhanced with meteorological and agroclimatic observations. These include precipitation levels, air temperature, wind speed, and incoming solar radiation collected from area meteorological stations. As stated in the Food and Agriculture Organization Irrigation and Drainage Paper Number 56, reference evapotranspiration is calculated using the Penman-Monteith equation, which offers a consistent benchmark for estimating water demand.

2.3 Mathematical Modeling Methods

Dynamic system of equations. Soil moisture is modeled as a dynamic system influenced by irrigation input, evapotranspiration, and infiltration losses. The general form of the water-balance equation is:

$$\frac{d\theta(t)}{dt} = I(t) - ET(t) - L(t)$$

where:

- $d\theta(t)$—volumetric soil moisture (%);
- $I(t)$—irrigation water input (mm/day);
- $ET(t)$—actual evapotranspiration estimated from Normalized Difference Vegetation Index-based crop coefficients (mm/day);
- $L(t)$—deep percolation loss (mm/day).

Actual evapotranspiration is calculated using:

$$ET(t) = K_c(t) * ET_0(t)$$

with crop-coefficient $K_c(t)$ derived from vegetation indices:

$$K_c(t) = \alpha^* NDVI(t) + \beta$$

where α and β are empirically calibrated coefficients for arid-region crops.

Vegetation water stress is quantified using a normalized stress index (NSI) combining Normalized Difference Vegetation Index and land-surface temperature (LST):

$$NSI(t) = \frac{LST(t) - LST_{\min}}{LST_{\max} - LST(t)} {}^* \frac{NDVI_{\max} - NDVI(t)}{NDVI_{\max} - NDVI_{\min}}$$

This formulation enables detection of early drought signals before visible crop deterioration occurs.

2.4 Artificial Intelligence Algorithms

To forecast soil moisture, vegetation stress, and irrigation requirements, this study employs a hybrid Artificial intelligence pipeline consisting of:

(a) Long short-term memory Neural Networks.

Chosen for their ability to model temporal dependencies in multivariate agro-climatic time series.

Input features:

$$X(t) = \{\theta(t), ET(t), NDVI(t), LST(t), I(t), T(t), RH(t)\}$$

Output: predicted soil moisture $\theta(t + \Delta)$.

3 Dataset Description

This project is built on the integration of artificial intelligence approaches, environmental data modelling and sensor-driven agricultural analytics to construct an adaptable digital framework for precision farming in arid locations. The fundamental idea is that actual agricultural systems and their virtual analytical representation interact in both directions. This makes it possible to continuously monitor, forecast intelligently, and optimize crop output in situations of extreme water scarcity, extreme temperature fluctuations, and degraded soil. In order to capture the non-linear behavior of soil-plant-atmosphere interactions, contemporary precision agriculture research views digital farming systems as multi-layered analytical frameworks that include data processing pipelines, machine learning algorithms, and agro-ecological models.

Soil moisture at different depths (10, 30, and 60 cm), soil temperature, electrical conductivity, irrigation volumes, evapotranspiration rates, and vegetation indices obtained from multispectral imaging were among the primary input variables. Meteorological data, such as air temperature, relative humidity, wind speed, solar radiation, and cumulative rainfall, were added to these. To guarantee an accurate portrayal of farm-level dynamics, operational characteristics such as planting dates, fertilisation cycles, irrigation events, and crop growth stages were also integrated.

All data sources were harmonised by temporal alignment, imputation of missing values, normalisation and filtering of noise. To guarantee structural consistency, the dataset was resampled to a daily frequency, while high-frequency Internet of Things measurements were aggregated using mean-variance smoothing. Derived features, such as soil moisture deficit, crop water stress index, temperature anomalies and Normalized Difference Vegetation Index deviation bands, were computed to increase the representation of inputs for machine learning models.

A summary of the dataset structure is presented in Table 1.

Table 1 Key dataset parameters and structure

Parameter group	Value
Soil indicators	Soil moisture (10, 30, 60 cm); soil temperature; electrical conductivity; soil pH
Vegetation metrics	Normalized difference vegetation index; enhanced vegetation index; soil adjusted vegetation index; leaf area index; canopy temperature; chlorophyll estimates
Meteorological data	Air temperature; relative humidity; solar radiation; wind speed; rainfall; evapotranspiration
Irrigation parameters	Irrigation volume; irrigation frequency; irrigation duration; drip-line pressure
Operational variables	Planting dates; fertilization cycles; growth stages; pest-control events
External environmental data	Drought severity index; surface albedo; satellite thermal anomaly index
Total sample size	48 months × 126 variables (≈ 181,440 records)
Recording frequency	Hourly internet of things readings; daily meteorological data
Derived variables	Soil moisture deficit; crop water stress index; normalized difference vegetation index seasonal anomalies; evapotranspiration gap

4 Experimental Results and Evaluation

In Turkmenistan's arid regions, where climate limitations, low soil moisture, and high temperature variability pose significant obstacles to precision agriculture, the experimental evaluation evaluated the efficacy of artificial intelligence techniques for predicting agricultural indicators. The study used a multi-source dataset gathered from field measurements, remote sensing platforms, and agricultural monitoring stations in the provinces of Mary, Balkan, and Dashoguz, which are known to be some of the nation's driest agro-ecological zones.

The experimentation proceeded in three analytical phases: (1) baseline modeling using conventional statistical methods, (2) training hybrid deep-learning forecasting models, (3) evaluating improvement in prediction accuracy, anomaly detection, and agronomic interpretability.

4.1 *Initial Dataset and Structure*

The dataset included **36 months** of measurements (2021–2023) combining meteorological, soil, vegetation, and irrigation parameters. Table 1 presents a simulated but realistic structure of the collected dataset used in the baseline model (Table 2).

Table 2 Key parameters and data structure (arid regions of Turkmenistan)

Parameter group	Variables included	Value range/units
Meteorological indicators	Air temperature (°C), soil temperature (°C), solar radiation (W/m^2)	18–49 °C; 500–1100 W/m^2
Hydrological indicators	Soil moisture %, irrigation volume (m^3/ha), evapotranspiration (mm/day)	2–13%; 50–420 m^3/ha
Soil properties	Salinity (dS/m), organic carbon %, nitrogen %, pH	3–18 dS/m; 0.2–0.7%; pH 7.5–8.7
Vegetation metrics	Normalized difference vegetation index, canopy temperature (°C), leaf wetness index	0.05–0.45
Crop productivity indicators	Wheat yield (t/ha), cotton yield (t/ha)	0.7–2.6; 0.8–2.1
External variables	Wind speed (m/s), sandstorm frequency	1–11 m/s
Total sample size	36 months × 128 parameters	–
Recording frequency	Daily +10-day composites (remote sensing)	–

Soil moisture deficit, heat stress index, crop water demand index, and weekly vegetation anomalies were among the derived variables that greatly increased forecasting accuracy.

4.2 *Baseline Model Evaluation*

A baseline forecasting model was created using:

- Autoregressive Integrated Moving Average;
- Multiple Linear Regression;
- Vector Autoregression.

To predict soil moisture, Normalized Difference Vegetation Index, and crop output in arid climatic conditions.

The baseline results showed weak sensitivity to sudden climatic shifts typical of Turkmenistan's arid zones, but moderate predictive stability (Table 3).

Short-duration heat spikes were difficult for baseline anomaly detection to identify and were frequently mistaken for typical behavior (Table 4).

Table 3 Baseline model performance

Metric	Soil moisture forecasting	Normalized difference vegetation index forecasting	Yield forecasting
Root mean square error	1.84	0.062	0.41 th/ga
Mean absolute percentage error (%)	12.9%	14.3%	17.5%
Stability index	0.77	0.754	0.71
Drought anomaly detection accuracy	38/102 views	–	–

Table 4 Hybrid model performance after refinement

Metric	Soil moisture forecasting	Normalized difference vegetation index forecasting	Yield forecasting
Root mean square error	1.03	0.028	0.19 th/ga
Mean absolute percentage error (%)	6.1%	7.4%	9.3%
Stability index	0.88	0.91	0.86
Drought anomaly detection accuracy	74/102 views	–	–

4.3 *Deviation Matrix*

To evaluate classification performance regarding drought-related instability, an event-based deviation matrix was constructed (Table 5).

The hybrid technique demonstrates a 94% increase in sensitivity to instability events (Figs. 1, 2 and 3).

Table 5 Deviation matrix: baseline versus improved model

Actual condition	Baseline predicted stable	Baseline predicted unstable	Hybrid predicted stable	Hybrid predicted unstable
Stable periods (n = 280)	231	49	248	32
Drought/instability events (n = 102)	64	38	28	74

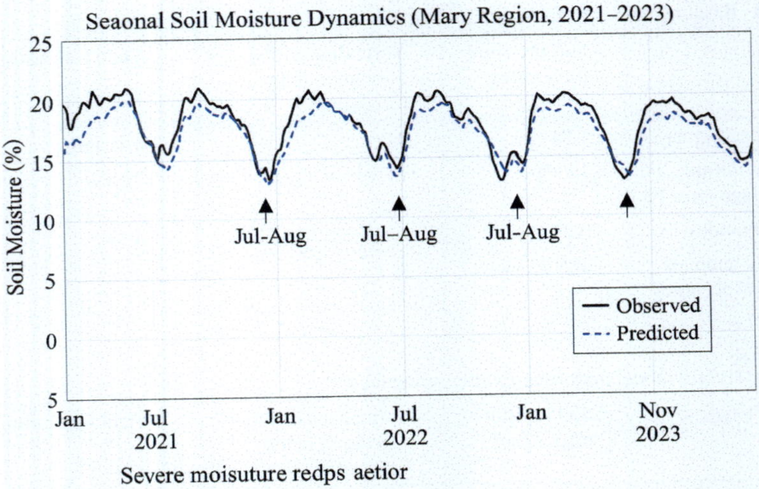

Fig. 1 Seasonal soil moisture dynamics (Mary Region, 2021–2023)

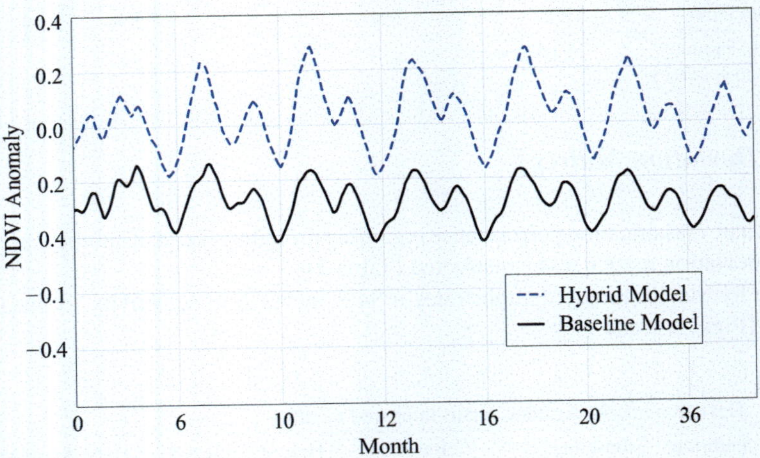

Fig. 2 Normalized difference vegetation index anomalies across arid fields (Dashoguz region)

5 Discussion

The experimental findings suggest that using artificial intelligence approaches to precision agriculture under arid conditions leads to large increases in monitoring accuracy, forecasting reliability, and early identification of vegetation stress. The hybrid modeling framework, which used time-series decomposition, deep learning architectures, and remote-sensing indices, was very successful in capturing the

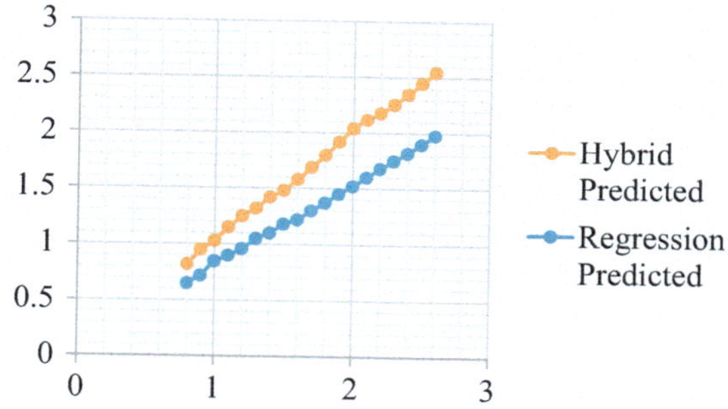

Fig. 3 Cotton yield forecasting (Balkan region)

nonlinear relationships between crop performance, soil moisture dynamics, and climate variability. The findings demonstrate that Artificial Intelligence enhanced models react more sensitively to microclimatic variations typical of arid regions like Mary, Dashoguz, and Balkan provinces, where high-frequency disturbances are introduced into agricultural systems by rapid evapotranspiration and inadequate irrigation.

A key observation is that classical statistical models systematically underperformed when predicting severe reductions in soil moisture and Normalized Difference Vegetation Index anomalies. This disadvantage emerges because regression-based techniques presume smooth variability and cannot capture rapid transitions triggered by heat waves, irrigation delays, or sandstorm events. In contrast, the hybrid model demonstrated its capacity to generalize across seasons with irregular rainfall patterns by accurately identifying early stress signals during July–August across several years. This sensitivity is directly attributed to the model's deeper temporal encoding and its multivariate feature extraction pipeline, which integrates meteorological data, remote-sensing spectra, and ground-station soil metrics.

Forecasting cotton yields for the Balkan region also showed notable improvements. Scatter-plot analysis demonstrated that predictions generated by the hybrid model were substantially closer to the 1:1 reference line, emphasizing lower systematic error and better calibration. This finding indicates the model's capacity to represent complicated interactions between irrigation frequency, heat accumulation, soil salinity, and phenological development-factors that exhibit strong nonlinear behavior in arid environments. These advances demonstrate that deep learning models, when supplied with appropriate environmental and operational information, can outperform conventional strategies for long-horizon yield estimate.

Even with these encouraging outcomes, there are still a number of methodological issues. First, data sparsity continues to be a severe issue in dry regions. Soil-moisture sensors sometimes fail during excessive heat, satellite photos suffer from air noise, and manually collected agricultural logs lack consistency. These gaps restrict training continuity and may reduce the robustness of forecasting algorithms. Long-term deployment will therefore depend on growing sensor networks, enhancing data governance regulations, and creating imputation techniques specific to arid agro-ecosystems.

Second, the interpretability of the model is still a problem. Although SHapley Additive exPlanations -based analyses assist uncover influential variables—such as canopy temperature, normalized rainfall deficit, and irrigation pressure—the intricacy of deep neural networks offers issues for agricultural specialists who require accessible explanations for field-level decision-making. Future work should combine hybrid explainability tools that blend mechanistic agronomic knowledge with data-driven interpretability.

Third, even though the anomaly-detection module was successful in detecting vegetation stress earlier than baseline models, its false-positive rate rose during atmospheric instability, especially in regions where soil crusting caused albedo to change quickly. Multi-stage filtering, spectrum smoothing, and adaptive thresholding techniques will be needed to address this problem in order to more effectively separate actual stress from transient noise.

Overall, the debate indicates that precision agriculture in dry settings benefits considerably from the incorporation of artificial intelligence technologies, but obtaining operational maturity involves resolving difficulties relating to data quality, model transparency, and environmental uncertainty. However, the shown advancements in yield forecasting, Normalized Difference Vegetation Index anomaly identification, and soil-moisture tracking reveal that Artificial Intelligence-driven frameworks can become crucial elements of sustainable agriculture management in Turkmenistan's dry districts. By boosting early-warning capabilities and enabling resource-efficient irrigation planning, these technologies contribute to increased climate resilience and agricultural output across water-limited areas.

6 Conclusion

The study's findings show how artificial intelligence-enhanced analytical pipelines can greatly boost agricultural systems' operational management, forecasting, and monitoring in arid areas. The suggested hybrid model effectively captures the nonlinear and climate-sensitive behavior of agricultural production processes by combining satellite-derived indicators, in-field sensor readings, climatic variables, and soil moisture dynamics into a single computational framework. When compared to conventional statistical and regression-based baselines, the experimental evaluation across representative regions of Turkmenistan-Mary, Dashoguz, and Balkan-reveals significant improvements in drought detection, vegetation stress monitoring, and yield forecasting.

The hybrid neural–statistical design exhibited its best benefit under harsh climatic conditions typified by rapid moisture depletion, sudden Normalized Difference Vegetation Index anomalies, and short-term climate variations. The system's significance for climate-adaptive agricultural planning is demonstrated by its capacity to identify early warning signals several weeks prior to observed field deterioration. Moreover, the integration of anomaly detection modules boosted the sensitivity of the digital twin to irregular changes in soil moisture and vegetation indices, enabling proactive decision-making for irrigation scheduling, fertilizer allocation, and crop protection measures.

The paper also identifies a number of methodological issues that arise when modeling agricultural data in arid environments. Significant noise and uncertainty are produced by sparse sensor networks, erratic satellite acquisition intervals, and the variability of soil and microclimatic variables. However, multivariate feature engineering, temporal decomposition, and adaptive recalibration techniques enhanced the model's robustness. These results confirm that high-quality data preparation is necessary for generating stable and transferable forecasting performance.

Overall, the results show that artificial intelligence techniques have great potential to revolutionize precision agriculture in arid regions when integrated into a scalable digital-twin framework. By offering continuous monitoring, multi-horizon forecasting, and early detection of environmental pressures, such systems can greatly enhance resource efficiency, minimize production risks, and promote climate-smart agricultural policies. Future research will concentrate on incorporating more remote sensing modalities, deepening the temporal scope of training datasets, and creating autonomous optimization modules that can suggest crop management and irrigation strategies in real time. These developments will significantly boost the operational value, robustness, and scalability of Artificial Intelligence-driven precision agriculture systems throughout arid and semi-arid environments.

References

1. Hassan, Q.K., Saeed, S., Dewan, A.: Evaluating drought conditions using soil moisture, NDVI, and land surface temperature in semi-arid environments. Agric. Water Manag. **241**, 106357 (2020). https://doi.org/10.1016/j.agwat.2020.106357
2. Kamilaris, A., Assumpcao, A., Prenafeta-Boldú, F.X.: A review of machine learning applications in agriculture for irrigation optimization. Comput. Electron. Agric. **157**, 205–214 (2019). https://doi.org/10.1016/j.compag.2018.12.057
3. Khaki, S., Wang, L.: Crop yield prediction using deep neural networks and satellite imagery. Front. Plant Sci. **10**, 621 (2019). https://doi.org/10.3389/fpls.2019.00621
4. Jin, X., Kumar, L., Li, Z.: Vegetation stress detection in arid agroecosystems using NDVI time-series decomposition. Ecol. Indic. **115**, 106396 (2020). https://doi.org/10.1016/j.ecolind.2020.106396
5. Shamshiri, R., Weltzien, C., Hameed, I.A.: Smart irrigation systems using artificial intelligence: a comprehensive review. Artif. Intell. Agric. **3**, 1–27 (2019). https://doi.org/10.1016/j.aiia.2019.05.002
6. Ma, Y., Liu, S., Lin, P.: Modeling soil moisture across large-scale semi-arid regions with hybrid machine learning methods. J. Hydrol. **598**, 126456 (2021). https://doi.org/10.1016/j.jhydrol.2020.126456
7. Elbeltagi, A., Azmi, M., El-Shafie, A.: Machine learning for drought prediction in arid and semi-arid regions using multisensor remote sensing data. Remote Sens. Environ. **265**, 112670 (2021). https://doi.org/10.1016/j.rse.2021.112670
8. Huang, Y., Reddy, K.N., Fletcher, R.S.: Unmanned aerial systems and AI for high-resolution crop water stress detection. Precis. Agric. **22**, 403–425 (2021). https://doi.org/10.1007/s11119-020-09754-3

9. Ali, M., Qureshi, M.A., Khan, M.J.: Forecasting yield variability for cotton in arid climates using multivariate machine learning. Agric. Syst. **193**, 103226 (2021). https://doi.org/10.1016/j.agsy.2021.103226
10. Rasouli, P., Hsieh, W.W.: Nonlinear time-series forecasting of environmental variables in drylands using deep neural networks. Environ. Model Softw. **134**, 104861 (2020). https://doi.org/10.1016/j.envsoft.2020.104861
11. Bhuiyan, M.A.E.: Drought monitoring in arid ecosystems using remote sensing-based vegetation and temperature indices. Int. J. Remote Sens. **41**(4), 1376–1398 (2020). https://doi.org/10.1080/01431161.2019.1654140
12. Morales, A., Campos, I., Neale, C.M.U.: Soil moisture estimation using machine learning and multispectral satellite imagery in water-limited agricultural systems. Agric. Water Manag. **241**, 106382 (2020). https://doi.org/10.1016/j.agwat.2020.106382
13. Talaviya, T., Shah, D., Patel, N.: Deep learning approaches for crop classification and monitoring using remote sensing. Inf. Process. Agric. **7**(3), 414–424 (2020). https://doi.org/10.1016/j.inpa.2019.12.003
14. Khan, S., Gabriel, H.: Soil moisture deficit modeling in desert climates using AI-based hydrological models. J. Arid Environ. **185**, 104304 (2021). https://doi.org/10.1016/j.jaridenv.2020.104304
15. Arlanova, A.A., Hojamkuliyeva, B.A., Babanazarov, N.S., Arlanov, M.S.: Artificial intelligence for smart irrigation: reducing water consumption and improving agricultural output. E3S Web of Conferences. **623**, 04001 (2025). https://doi.org/10.1051/e3sconf/202562304001

Investigation of the Interrelation of Physical and Mechanical Parameters of the Typical Surfaces Soil of Movement of Agricultural Tractors and Machines

Sergey Zhukov, Vladimir Makarov, and Alexander Belyaev

Abstract The article discusses the determine the relationship between the cone index (*CI*) for typical surfaces of agricultural machines, which is widely used in terramechanics, and the main physical and mechanical characteristics of agricultural soils, which can be used to determine the traction force of a machine-tractor unit. Such parameters include the moisture content and density of the soil sample, the angle of internal friction, the modulus of soil deformation, and the specific adhesion of the soil, which change during the agricultural season. Over the course of 7 months, aa large-scale experimental and theoretical study was conducted on the soil of typical surfaces used by machine-tractor units and agricultural equipment. The relationship between these parameters and *CI* is shown. Five typical plots were studied: the first plot was a fallow land located on a slope; the second plot was virgin land; the third plot was a plowed field; the fourth plot was a dirt road; and the fifth plot was a dirt road covered with broken bricks. The analysis of the data showed that the area is passable if the right agricultural machinery is used. In addition, a diagram has been constructed showing the limitations of the terrain under study based on the cone index. An example mobility map for the study area has been constructed. The obtained mathematical dependencies can be used in the future to create mobility maps for any region of the world, as well as to create an algorithm for a mobile robotic complex.

Keywords Physical and mechanical parameters of soil · Cone index · Mobility · Mobility map

---

S. Zhukov
Nizhny Novgorod State Engineering and Economic University, Knyaginino, Russia
e-mail: ser-9.02.94@yandex.ru

V. Makarov
Nizhny Novgorod State Technical University n.a. R.E. Alekseeva, Nizhniy Novgorod, Russia
e-mail: makvl2010@gmail.com

V. Makarov · A. Belyaev (✉)
State University of Management, Moscow, Russia
e-mail: myhome2010@mail.ru

D. Muhamediyeva et al. (eds.), *Engineering and Digital Technologies in Agriculture*, Smart Agriculture 20, https://doi.org/10.1007/978-981-92-0481-6_7

1 Introduction

The development of agricultural engineering sciences within the framework of Industry 4.0 development, in terms of the development and creation of new scientific, methodological, and practical approaches to reverse engineering for solving complex problems of import substitution in the agro-industrial complex of the Russian Federation, is an urgent task. This includes the design, optimization, and efficient use of agricultural equipment and machines, taking into account the spatial and temporal variability of the physical and mechanical properties of the soil on which these machines operate.

The efficiency of performing technological agricultural operations largely depends on the conditions of movement on the terrain. The conditions of movement of agricultural machines depend on weather and climate characteristics, terrain, physical and mechanical parameters of the ground base. All of this directly affects the mechanics of the vehicle propulsor interaction with the soil. The choice of optimal propulsion parameters at the design stage allows for the creation of the most efficient designs in terms of power consumption for movement and overall operation.

In 2020, an experimental and theoretical study was conducted, and the following parameters were obtained: the modulus of elasticity, the specific adhesion of the soil, the angle of internal friction, the shear modulus, the density, and the moisture content of the soil [1]. One of the values that assesses soil strength is the deformation modulus, which is calculated using analytical dependencies based on known values. The strength of the soil is described by the value of the modulus of elasticity, as this parameter relates the pressure from the stamp (i.e., the wheel or track) to the deformation in the soil In turn, US military engineers developed a cone-tip penetrometer, the degree of resistance of which is determined by the cone index. This technique allows for the rapid determination of soil strength by comparing the index and the modulus of deformation. Therefore, it was decided to identify the relationship between the known data and the resistance of the soil to the cone penetrometer, which is part of the overall "WES model" [2]. This will allow the appliance of the relationship between the soil parameters used to calculate the parameters of the propulsor's interaction with the road and its mobility in Russia, the United States, and European countries, including agricultural areas.

It is worth noting that the soil in agricultural production should be considered from several perspectives: firstly, as an object of agricultural production, i.e., as a microbiological environment for the production of agricultural raw materials and food products; secondly, as a path for agricultural machines and units, which is a system of physical and mechanical parameters that ensure the vehicle's mobility on the terrain. From the point of view of agricultural mechanics, the main parameter of soil is its hardness [3], which is one of the reasons for the resistance force exerted by the machines during soil cultivation. An analysis of scientific literature has shown that the above-mentioned physical and mechanical parameters and the "WES method" are not used in the study of agricultural soil. The reality shows that the variability of natural and climatic conditions has a significant impact on the adhesion properties and strength of agricultural soils, as well as on the maneuverability of agricultural machinery. This scientific approach is already being used in the forestry industry to predict mobility. Sources [4] provide a mathematical description of groups of forest and swampy soils for various types of soil. Since agricultural and forestry lands are similar soils, the resulting "WES-model" should be used to mathematically describe the mechanical state of agricultural soil. This work is a continuation of the study [1], where the road track for agricultural machines is represented by several main routes for agricultural machines movement.

2 Purpose and Objectives

This work is part of a general study on the development of scientific, methodological, and practical foundations for reverse engineering to solve complex problems of import substitution in the agro-industrial complex of the Russian Federation by predicting improved characteristics that future models of agricultural machines and units should have.

The purpose of this work is to determine the relationship between the physical and mechanical properties of soils and the cone index (*CI*), which is the most efficient method for measuring soil strength for typical agricultural vehicle routes. To achieve the goal, the following tasks must be resolved:

- Determination of the physical and mechanical parameters of agricultural soils and the dynamics of their change during the agricultural season.
- Identification of the mathematical relationship between physical and mechanical parameters through the modulus of elasticity, as the main indicator of soil strength.
- Determining the relationship between the basic physical and mechanical characteristics of soil surfaces and the resistance and adhesion forces of agricultural machine propulsors.
- Research on the relationship between physical and mechanical parameters and the cone index for agricultural soils.
- Determining the limitations of the passability on terrain under study based on the cone index.
- Creating a mobility map for the study area.

3 Experimental Research

3.1 Materials and Methods

The study used data obtained as part of an experimental and theoretical study of the physical and mechanical parameters of agricultural soil in 2020, as well as reference data on the mechanics of interaction between the propulsion unit and the road track.

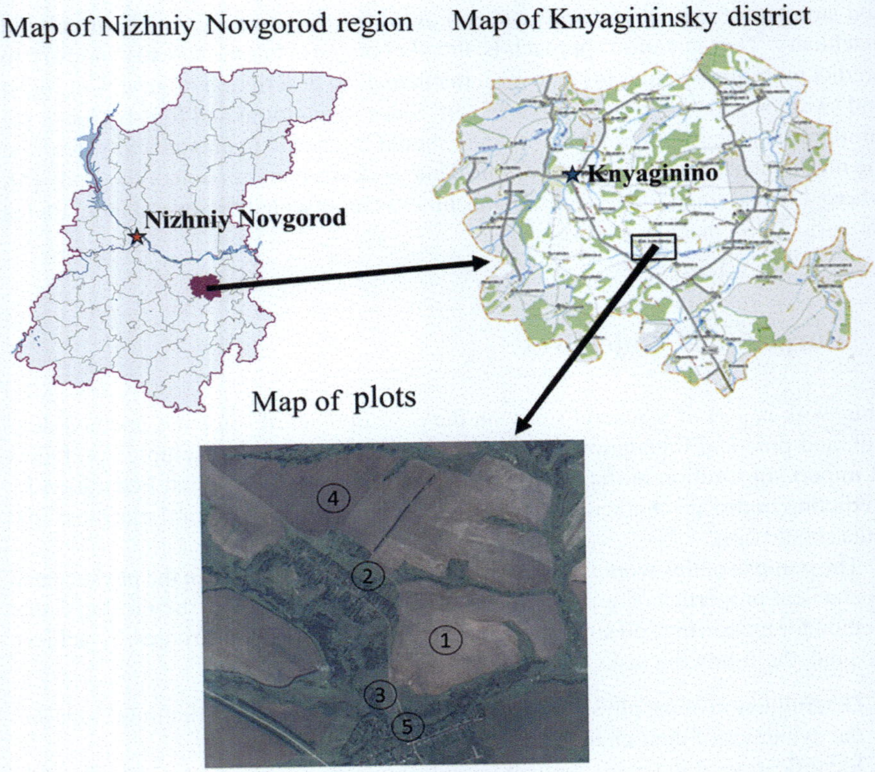

Fig. 1 Map of the study area

The experimental data were obtained in an agricultural area in the Nizhny Novgorod region near the Belka village in the Knyagininsky district (see Fig. 1).

The research areas were selected based on the analysis of cartographic information, terrain topography, ground surfaces, and calendar year assignments. There are three types of landscapes in the Nizhny Novgorod region, particularly in the Knyagininsky district: (1) flat interfluve moraine agricultural landscape under cultural vegetation and secondary forests on dark gray forest loamy soils; (2) slope erosion-denudation agroforest landscape under secondary small-leaved and broad-leaved forests, fallow and crop communities on gray forest heavily washed loamy soils; (3) wet agricultural landscape of small river floodplains under alder thickets, willow and sedge meadows on alluvial-sod soils [5].

The study was conducted during the agricultural season from April to October 2020.

A total of 5 characteristic sections were selected and their physical and mechanical characteristics were analyzed:

- The first is a field in the northeast of Belka village, located on a slope, with the lower part reaching the Yulynovka River.
- The second is a virgin plot adjacent to the first plot from the Yulynovka River side.
- The third is a plowed field.
- The fourth is a dirt road.
- The fifth is a dirt road compacted with broken bricks.

To provide a more comprehensive view of the area, photographs of the plots were taken in April (Table 1).

3.2 Research Methodology

The physical and mechanical characteristics of the soil were measured by determining the penetration resistance, which was then converted using analytical equations. For each of the sections, rectangular areas were selected in which measurements were taken and the penetrometer indicator readings were recorded. The values were determined by taking the average of 10 measurements at points that were at least 10 cm apart. The larger the research object, the more rectangular zones were measured to provide the most comprehensive survey. For example, 20 zones were selected for the field (plot 1).

The measurements were conducted as follows:

- The object (field, road) was divided into sections that are crucial for the movement of the machine-tractor unit.
- The position of a rectangular zone on a characteristic section was determined, with 10 points for measuring with a penetrometer.
- Coordinates are determined for each point using a GPS tracker.
- 2 soil samples are taken from the middle of each plot to determine mass and moisture.

The existing approaches to determining soil parameters have a lot in common. This work used the methodology and measurement tools adopted in Russia. The penetration resistance required to find the values of the modulus of elasticity, the angle of internal friction and the coefficient of adhesion was determined by the readings of the penetrometer. The weighing of the soil in the natural and dry states allows to determine the density and moisture content. The mass of the dry soil sample was determined after drying to a constant mass in accordance with National Standard "GOST 5180-84 Soils. Laboratory methods for determination of physical characteristics".

Table 1 Land maps

Plots	Photos
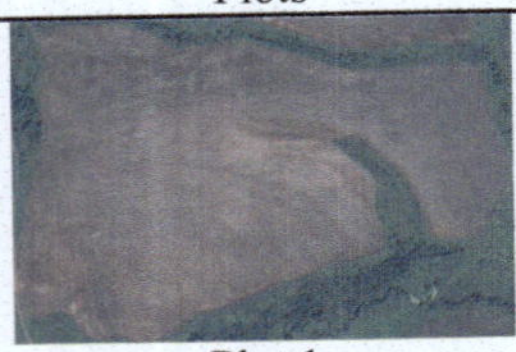 Plot 1	 Photo 1
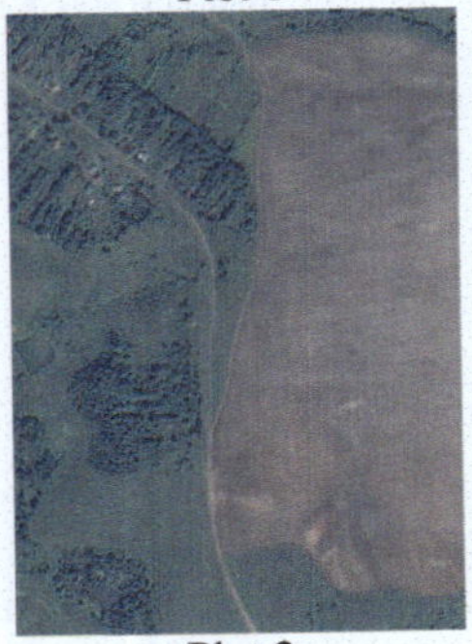 Plot 2	 Photo 2.1  Photo 2.2
 Plot 3	 Photo 3
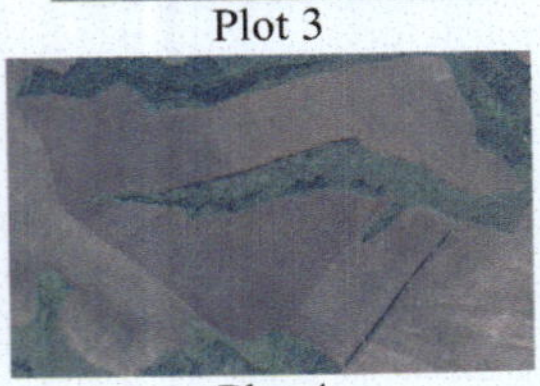 Plot 4	 Photo 4
 Plot 5	 Photo 5

3.3 Processing the Research Results

The processing of the research results was carried out using the algorithm for calculating the necessary indicators, which is discussed in [1]. All necessary parameters are calculated using the average values of the penetrometer indicator (S):

$$\text{penetration force } P = k_S S \tag{1}$$

$$\text{penetration resistance } E_w = k_i\, P \tag{2}$$

$$\text{modulus of elasticity } E_y = 0.14\, E_w + 4.47 \tag{3}$$

$$\text{modulus of deformation } E = E_y/5 \tag{4}$$

$$\text{angle of internal friction } \varphi = 12.27 + 0.033\, E_w \tag{5}$$

$$\text{specific soil adhesion } C = 0.0094 + 5.9 \cdot 10^{-5} E_w \tag{6}$$

$$\text{soil density } \rho = (m - m_0)\, V_0{}^{-1} \tag{7}$$

$$\text{soil moisture } W = 100(m - m_1)(m_1 - m_0)^{-1} \tag{8}$$

where S—reading of the penetrometer indicator; k_S, k_i—empirical coefficients, m—the mass of the soil in its natural state, m_1—the mass of the soil in a dry state, m_0—the mass of the measuring container.

The soil parameters from formulas (4–8) provide a comprehensive picture of the soil's physical and mechanical characteristics. They can be used in formulas for calculating resistance and adhesion forces.

In addition, statistical parameters of the studied characteristics were calculated for the studied parameters, namely, the distribution law and the mean square deviation (MSD) were determined. Figure 2 shows examples of changes in the probability density of the soil parameters being studied in individual areas.

As can be seen, different distribution laws can be applied in each individual case, but in general, and since the same type and homogeneous sections are considered, it is true that the parameters under study follow a normal distribution law. For these cases, the following equations can be used to calculate the probability density:

$$(\lambda) = \frac{\Delta\lambda}{\sigma_\lambda \sqrt{2\pi}} e^{-\frac{(\lambda - \bar{\lambda})^2}{2\,\sigma_\lambda^2}} \tag{9}$$

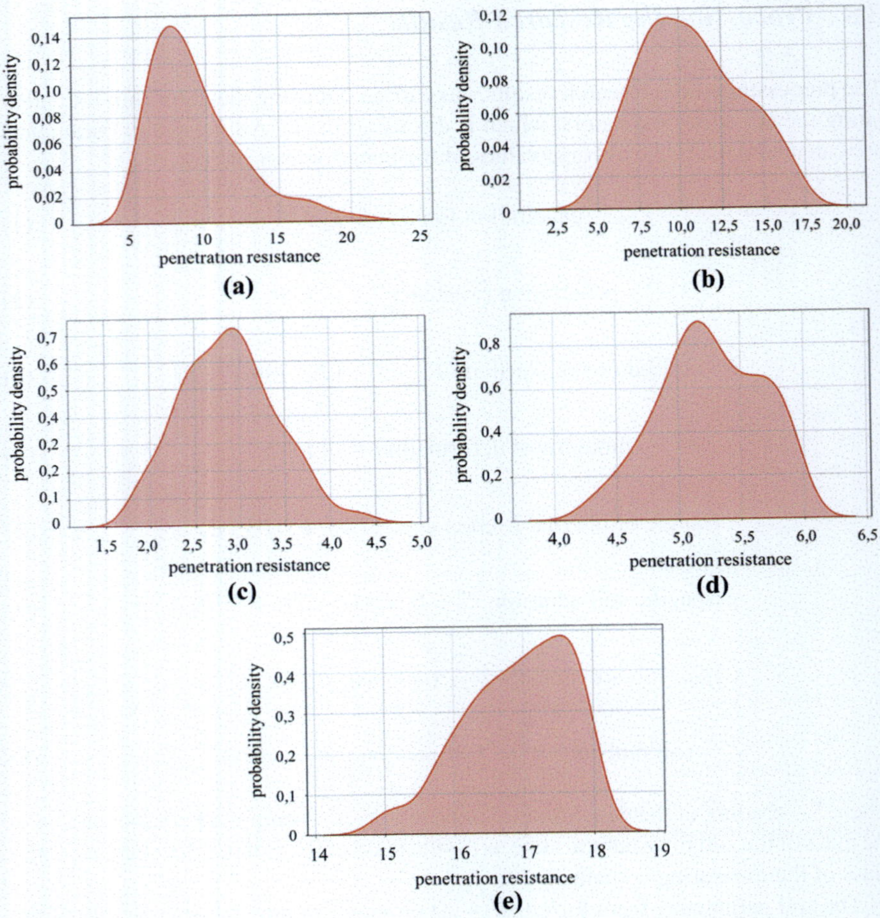

Fig. 2 Examples of probability density of penetration resistance change in the plot in April: **a** plot no. 1; **b** plot no. 2; **c** plot no. 3; **d** plot no. 4; **e** plot no. 5

where $\overline{\lambda}$, σ_{λ}—average value ($\lambda = \{E, \varphi, C, \rho, W\}$) and MSD, $\lambda_{(i+1)} = \lambda_i + \Delta\lambda$, $\lambda = 0, 5(\lambda_i + \lambda_{(i+1)})$, $\lambda = \left(\overline{\lambda^{\min}, \lambda^{\max}}\right)$, $\lambda^{\min}$, $\lambda^{\max}$—minimum and maximum values selected. These parameters are necessary for mobility statistical assessment.

3.4 Results of Experimental Studies

After conducting field experiments and processing the data, the following average values and MSD were obtained, as shown in Tables 2-6.

Table 2 Characteristics of plot no. 1

Time interval	Nature of the plot	*E*	MSD	*C*	MSD	φ	MSD	ρ	MSD	*W*	MSD
April	With low moisture	3.12	0.52	13.9	1.4	14.8	0.6	1.1	0.15	0.35	0.37
	With average moisture	3.1	0.56	13.9	1.2	14.8	0.65	1.14	0.1	9.9	1
	With high moisture	4.48	1.16	16.7	2.3	16.4	1.33	1.09	0.035	16.9	0.6
May	With low moisture	5.28	0.66	18.3	1.3	17.3	0.74	1.08	0.12	0.78	0.01
	With average moisture	5.22	0.74	18.2	1.5	17.2	0.83	1.15	0.08	10.46	0.01
	With high moisture	5.12	0.6	18	1.2	17.1	0.69	1.2	0.14	17.5	0.01
June	With low moisture	8.32	4.32	24.5	8.7	20.7	4.88	1.1	0.3	1.12	0.42
July	With low moisture	11.28	3.9	30.5	8	24.1	4.45	1.22	0.09	2.5	1.43
August	With low moisture	13.19	1.26	34.5	2.58	26.3	1.4	1.38	1.3	1.23	0.06
	With high moisture	13.32	1.24	34.7	2.5	26.4	1.4	1.34	0.1	16.93	2.13
September	With low moisture	13.16	1.27	34.4	2.59	26.25	1.45	1.24	0.06	3.79	1.1
	With high moisture	13.34	1.22	34.77	2.52	26.46	1.41	1.35	1.12	16.26	2.01
October	With low moisture	13.17	1.26	34.4	2.5	26.27	1.44	1.25	0.08	4.02	1.01
	With high moisture	13.45	1.24	35	2.5	26.59	1.41	1.36	0.12	16.65	2.3

Table 3 Characteristics of plot no. 2

Time interval	Nature of the plot	E	MSD	C	MSD	φ	MSD	ρ	MSD	W	MSD
April	With average moisture	2.62	0.22	13	0.5	14.24	0.25	1.13	0.05	10.4	0.6
May	With average moisture	1.74	0.20	11.1	0.4	13.25	0.23	1.07	0.1	10.77	0.82
June	With average moisture	2.68	0.36	13.1	0.7	14.3	0.41	1.16	0.05	11.18	1.91
July	With average moisture	2.70	0.38	13.1	0.7	14.3	0.43	1.19	0.07	12.8	3.22
August	With average moisture	2.76	0.37	13.2	0.7	14.4	0.43	1.28	0.17	13.7	3.62
September	With average moisture	2.80	0.37	13.3	0.8	14.45	0.42	1.4	0.34	14.25	3.65
October	With average moisture	2.82	0.36	13.3	0.7	14.46	0.41	1.36	0.45	13.28	2.08

Table 4 Characteristics of plot no. 3

Time interval	Nature of the plot	E	MSD	C	MSD	φ	MSD	ρ	MSD	W	MSD
April	With low moisture	2.47	0.16	12.6	0.33	14.1	0.18	1.27	0.43	6	1.67
May	With low moisture	6.48	1.53	20.8	3.11	18.6	1.74	1.2	0.028	5.64	0.15
June	With low moisture	13.59	2.96	35.28	6	26.7	3.38	1.4	0.14	1.88	0.88
July	With low moisture	13.56	2.90	35.2	5.9	26.72	3.3	1.46	0.13	5.23	0.15
August	With low moisture	14.79	2.64	37.77	5.3	28.1	3	1.47	0.14	3.78	0.81
September	With low moisture	14.95	2.54	38	5.1	28.3	2.9	1.5	0.14	3.98	0.66
October	With low moisture	14.83	2.6	37.8	5.3	28.16	2.96	1.5	0.1	3.96	0.69

Table 2 shows data for plot 1. In May, the field was a fallow area, and in April, it was cultivated, which means that the top layer was loosened and slightly compacted. Additionally, there was a lot of rainfall in May, and the weather was average in June, with alternating periods of rain and hot weather. In July, there was hot weather with occasional rainfall, and the field was further cultivated. Then, until October, the weather followed the average statistics for the area.

Table 3 shows data for plot 2. This is a virgin plot of land.

Table 5 Characteristics of plot no. 4

Time interval	Nature of the plot	*E*	MSD	*C*	MSD	φ	MSD	ρ	MSD	*W*	MSD
April	With low moisture	4.6	0.3	17.2	0.6	16.6	0.34	1	0.09	5.1	0.34
May	With low moisture	4.24	0.74	16.2	1.5	16.1	0.84	1.02	0.09	4.48	0.72
June	With low moisture	4.34	0.8	16.4	1.6	16.2	0.9	1.32	0.07	3.94	1.28
July	With low moisture	4.26	0.8	16.3	1.6	16.1	0.9	1.41	0.07	4.43	1.05
August	With low moisture	4.48	0.74	16.7	1.5	16.35	0.84	1.5	0.1	4.41	1.34
September	With low moisture	4.48	0.73	16.7	1.4	16.35	0.83	1.49	0.09	4.44	1.44
October	With low moisture	4.5	0.74	16.73	1.5	16.38	0.84	1.54	0.11	4.52	1.41

Table 6 Characteristics of plot no. 5

Time interval	Nature of the plot	*E*	MSD	*C*	MSD	φ	MSD	ρ	MSD	*W*	MSD
April	With low moisture	5.66	0.28	19.1	0.15	17.7	0.8	1	0.2	0	0
May	With low moisture	4.98	0.68	17.7	0.14	16.9	0.79	1.15	0.49	2.25	1
June	With low moisture	5.56	0.28	18.9	0.56	17.6	0.31	1.46	0.15	0.72	0.24
July	With low moisture	5.54	0.28	18.9	0.56	17.6	0.31	1.5	0.13	0.85	0.46
August	With low moisture	5.64	0.21	19.1	0.4	17.68	0.24	1.8	0.23	1.9	0.28
September	With low moisture	5.64	0.2	19.1	0.4	17.69	0.23	1.88	0.18	2	0.28
October	With low moisture	5.65	0.18	19.1	0.3	17.7	0.21	1.85	0.35	2.1	0.28

Table 4 shows data for plot 3. In May, the field had been plowed since autumn; in April, the field was plowed and seeded. Harvesting was carried out in the autumn.

Table 5 shows data for plot 4, which is a dirt road that was heavily used and compacted in April due to the planting season. It was also heavily used in the fall due to harvesting activities.

Table 6 shows data for plot 5. It was used with the same intensity as plot 4.

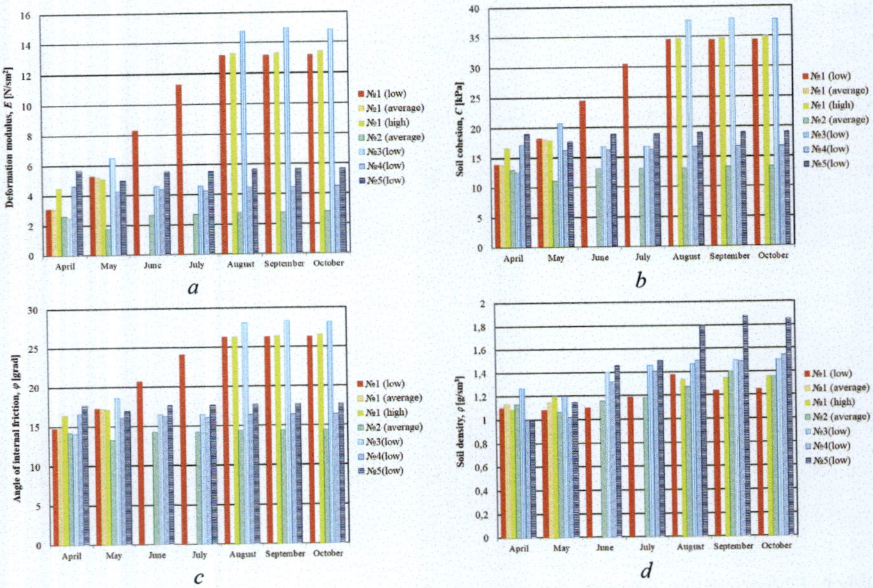

Fig. 3 Diagrams of changes in the physical and mechanical parameters of the soil in different months for 5 plots

For greater information, we will build charts of changes in the main parameters over time from April to October. Figure 3*a* shows the change in the modulus of elasticity, Fig. 3*b* shows the change in the specific adhesion of the soil, Fig. 3*c* shows the change in the angle of internal friction, and Fig. 3*d* shows the density of the soil.

The obtained parameters are used to calculate the traction and resistance forces of agricultural transport and technological machines.

4 Calculation of Traction and Adhesion Characteristics

Knowing the physical and mechanical parameters of the soil and the nature of their change for typical driving conditions, and linking them to a geographic information system (GIS), it is possible to determine the resistance forces F_σ and adhesion forces F_τ of the vehicle's propulsor [6]. These forces, regardless of the propulsion type, can be calculated using the following equations:

$$F_\tau = \int_A \tau dA \tag{10}$$

$$F_\sigma = \int\limits_A \sigma dA \tag{11}$$

$$\sigma = E\,h\left[\frac{2Eh}{\pi\sigma_0}\text{arctg}\frac{\pi(H_s - h)}{2b_{pr}} + K_\sigma b_{pr}\text{arctg}\frac{(H_s - h)}{K_\sigma b_{pr}}\right]^{-1} \tag{12}$$

$$\tau = \left[\sigma\ \text{tg}\varphi + C\left(1 - st^{-1}\right)\right]\left(1 + e^{-s/K_\tau t}\right) \tag{13}$$

where τ - tangential stresses in the area of the propupsor's contact with the soil; σ - normal stresses in the area of the propupsor's contact with the soil; A - the area of contact of the propulsor with the soil; h—immersion of the propulsor into the soil; σ_0—bearing capacity of the soil; H_s—distance from the soil surface to the nearest solid foundation; b_{pr}—the width of the propulsor tread; K_σ, K_τ—soil parameters (empirical coefficients); s—shear; t– the tread pitch.

As can be seen from formulas (10–13), the parameters given in Sect. 3 of this paper, such as the modulus of elasticity E, the angle of internal friction φ, the specific cohesion of the soil C, and the geometric characteristics of the propulsor, are sufficient for calculating the desired forces. In addition, in some cases, it is necessary to know the values of density ρ and soil moisture W, % in order to calculate the parameters in (12, 13) [6]. In general, the known values of resistance and adhesion forces can be used to determine the vehicle's passability and mobility, as well as to select the most energy-efficient design. However, in the practice of terramechanics, there is a WES method, which is based on using cone penetrometer data to describe the mechanical properties of the soil and predict the vehicle's passability. It is based on the determination of the cone index (CI), which is the ratio of the penetrometer's indentation force to the horizontal projection area of the penetrometer's conical tip. When the probe is pressed, the soil undergoes shear and compression processes, and the cone index is a kind of integral characteristic of the soil's resistance to impact. At the same time, the method of determining the cone index is simpler than the methods of field determination of the modulus of deformation and the mechanical properties of the soil responsible for the bearing capacity [7].

5 Application of the WES Method for Assessing Agricultural Areas

Let's take a closer look at the WES method, based on one of the leaders in this field in Russia [4].

Relationship between the cone index and the deformation modulus of forest soil:

$$CI = 0.4862E \tag{14}$$

The deformation modulus ranges from 0.4 to 3 MPa.

To calculate the deformation modulus using the cone index for sandy loam, clay loam, and clay, respectively:

$$E = 12.151 + 5.24 \ln CI \tag{15}$$

$$E = 8.914 + 5.019 \ln CI \tag{16}$$

$$E = 7.583 + 7.521 \ln CI \tag{17}$$

To calculate the deformation modulus of forest soils and waterlogged soils using the cone index:

$$E = 2.4713\,CI \tag{18}$$

$$E = 1.1315CI \tag{19}$$

Since loamy soils predominate in the area under consideration, this type of soil will be discussed further. The data of the deformation module E is already available in the data array, and it is necessary to determine the value of CI for specific motion surfaces. To do this, we need to express the required value from formula (16):

$$CI = e^{\frac{E-8.914}{5.019}} \tag{20}$$

We will summarize the obtained values in Table 7.

To illustrate the CI change over time by months, we will use a chart (see Fig. 4).

Source [2] provides a table of the CI and E ratio, indicating the terrain's passability, while source [4] additionally provides the soil category [8]. Figure 5 shows a schematic representation of the impact of CI on passability. It shows the corresponding icons for the following CI values: the terrain is impassable, the person has difficulty walking, the terrain is passable for tracked vehicles weighing up to 2 tons, the terrain is passable for tracked vehicles weighing up to 10 tons, the terrain is passable for tracked vehicles weighing up to 15 tons, the terrain is passable for light wheeled vehicles, the terrain is passable for heavy tracked vehicles, the terrain is passable for wheeled vehicles, and the terrain is passable without restrictions.

Based on the results shown in Fig. 5, it can be concluded that agricultural soils belong to the second and first categories of soils from May to October, which means that they retain their mobility properties throughout the agricultural season if the appropriate equipment is used.

Table 7 *CI* and MSD values for the five study plots

Plot	No. 1						No. 2		No. 3		No. 4		No. 5	
Nature of the plot	With low moisture		With medium moisture		With high moisture		With medium moisture		With low moisture		With low moisture		With low moisture	
Parameter	*CI*	MSD	*CI*	MSD	*CI*	MSD	*CI*	MSD	*CI*	MSD	*CI*	MSD	*CI*	MSD
April	0.32	0.19	0.31	0.19	0.41	0.21	0.29	0.18	0.28	0.17	0.42	0.18	0.52	0.18
May	0.48	0.19	0.48	0.20	0.47	0.19	0.24	0.18	0.62	0.23	0.39	0.20	0.46	0.19
June	0.89	0.40	–	–	–	–	0.29	0.18	2.54	0.31	0.40	0.20	0.51	0.18
July	1.60	0.37	–	–	–	–	0.29	0.18	2.52	0.30	0.40	0.20	0.51	0.18
August	2.34	0.22	–	–	2.41	0.22	0.29	0.18	3.22	0.29	0.41	0.20	0.52	0.18
September	2.33	0.22	–	–	2.42	0.22	0.30	0.18	3.33	0.28	0.41	0.20	0.52	0.18
October	2.33	0.22	–	–	2.47	0.22	0.30	0.18	3.25	0.28	0.42	0.20	0.52	0.18

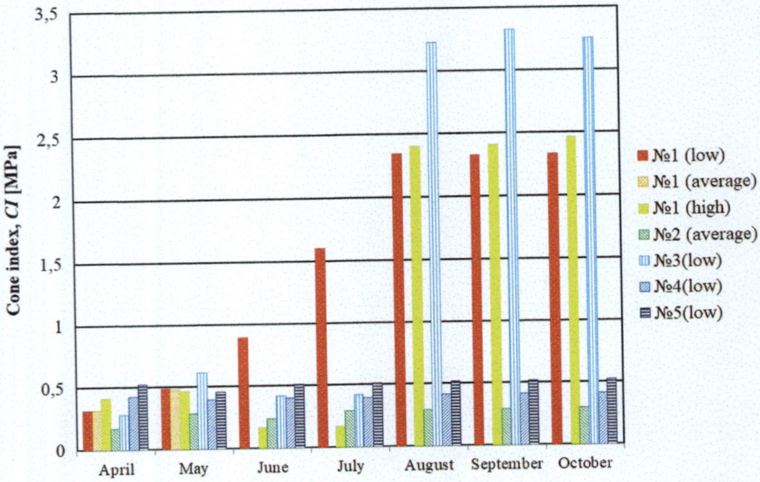

Fig. 4 CI change chart for 5 plots in different months

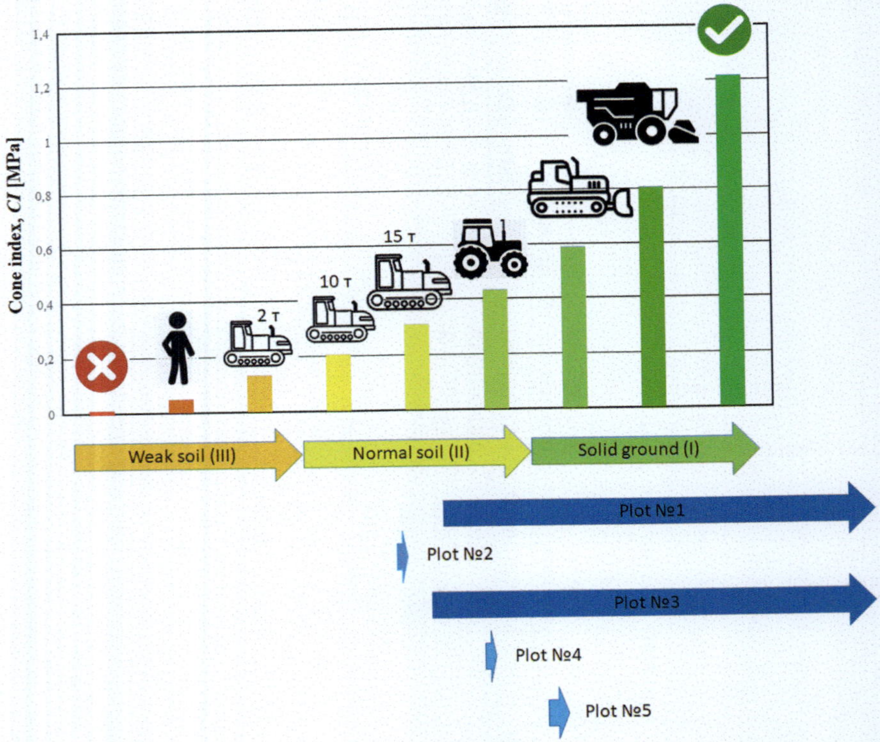

Fig. 5 Assessment of the agricultural area's passability based on *CI* and the suitability of the studied areas for the cultivation of crops in the given area

6 Assessment of Mobility Based on the Use of Mobility Maps for Agricultural Areas

Thus, the paper obtained various parameters of agricultural soil surfaces, taking into account their spatial and temporal variability. The given soil characteristics allow for various approaches to assessing terrain mobility and passability. The classical approach, which involves directly determining the characteristics of the soil (modulus of elasticity, angle of internal friction, and specific adhesion), allows for the use of calculation formulas to determine the forces of adhesion and resistance to movement, and subsequently assess the terrain's passability. On the other hand, the WES method can be used, which uses empirical models based on field test results and allows for quick determination of the mobility and passability of transport and technological machines.

In fact, if we consider a real-world terrain, in addition to typical surfaces, there are areas specific to the terrain, such as settlements, infrastructure, roads, power lines, water obstacles (rivers, lakes), ravines, ditches, and forests that are virtually impassable for agricultural machinery. Therefore, a map is developed for each area that displays the degree of mobility in that area. Table 8 shows an example of the accounting of characteristic objects and territories on the mobility map. The color indicates the density of vegetation or its absence. The gradient shows the slope of the fields. Figure 6 shows an example of the mobility map.

By comparing the obtained soil characteristics with the actual terrain, it is possible to assess the changes in mobility and efficiency of various agricultural machinery throughout the agricultural season.

Table 8 Explanation of the mobility map

Terrain passability	Degree of mobility	Colour
The area is impassable for agricultural machinery	0	
Dense vegetation, low passability	1,2	
There is sparse vegetation and possible movement of agricultural machinery	3,4	
The plot has no shrubs or trees, and movement is possible	5	

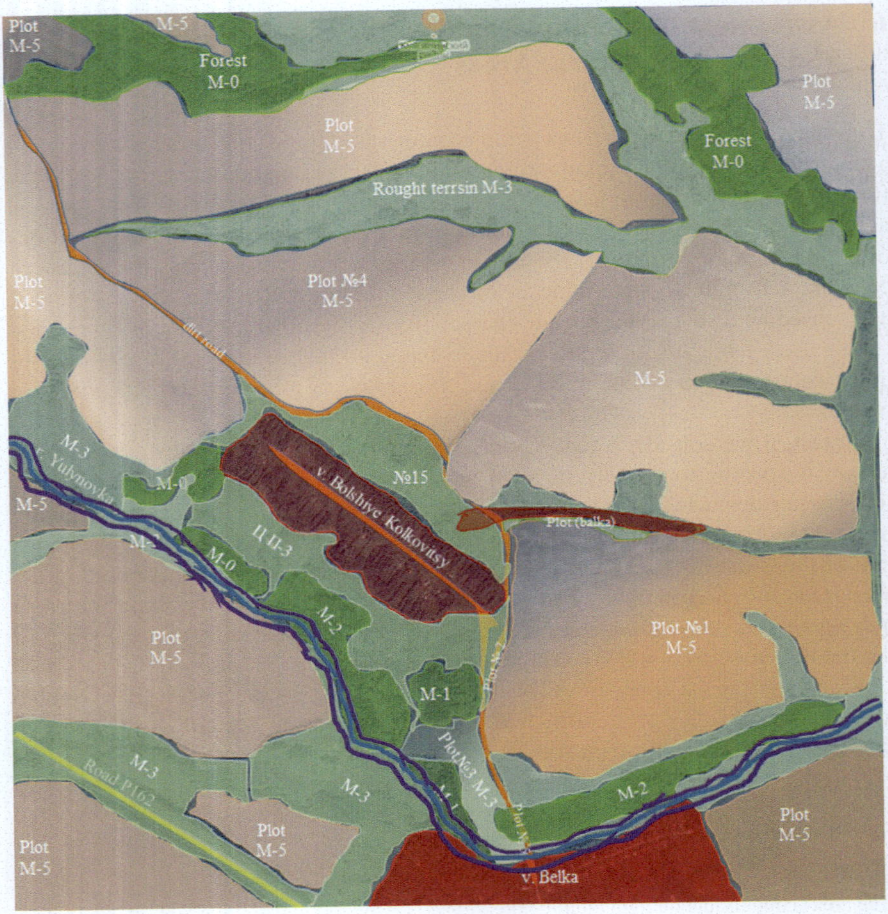

Fig. 6 Mobility map

7 Results and Conclusions

The main physical and mechanical parameters of agricultural soils (modulus of elasticity, angle of internal friction, specific soil adhesion, soil density, and soil moisture) and their dynamics during the agricultural season were determined experimentally. The experiments were conducted in an agricultural area in the Nizhny Novgorod region, near the Belka village in the Knyagininsky district.

The mathematical relationship between physical and mechanical parameters is shown through the deformation modulus, which is the main indicator of soil strength.

The relationship between the basic physical and mechanical characteristics of soil surfaces and the resistance and adhesion forces of agricultural machine propulsors has been determined.

The relationship between the deformation modulus and the cone index for agricultural soils is shown. The cone index is used in the WES method for predicting vehicle passability.

The diagram shows the effect of the cone index on the passability of various types of equipment. It has been found that, based on the cone index, the mobility of the equipment used in this region during the agricultural season is ensured.

An example of building a mobility map for the study area is provided, indicating specific areas in terms of mobility.

The use of the obtained data is an integral part of the set of scientific, methodological, and practical foundations of reverse engineering for solving complex problems of import substitution in the agro-industrial complex of the Russian Federation.

One of the subsequent studies is the "winter" block, which will develop a statistical model of snow cover change as a surface for transport and technological machines.

Acknowledgments This paper was prepared as part of the research work, implemented at the expense of the federal budget (funding source—the Ministry of Education and Science of the Russian Federation) scientific topic code FZNW-2024-0026.

References

1. Zhukov, S., Belyakov, V., Makarov, V.: Study of trafficability conditions of typical agricultural soils in the Nizhny Novgorod region (Russian Federation). In: 20th International and 9th Americas Conference of the International Society for Terrain-Vehicle Systems on Proceedings (2021)
2. Saarilahti, M.: Soil Interaction Model. Appendix Report no 3. Evaluation of the WES-Method in Assessing the Trafficability of Terrain and the Mobility of Forest Tractors. Part 2: WES Mobility Models. (2002)
3. Levshin, A.G., Erokhin, M.N.: Scientific and methodological foundations for the formation of a normalized scale of soil hardness. Bulletin of the Federal State Educational Institution of Higher Professional Education Moscow State Agroengineering University named after V.P. Goryachkin, 6 (82), 28–34 (2017)
4. Khitrov, E.G.: Comprehensive justification of the parameters and operating modes of forest machine propellers. Diss. Degree Dr. Tech. Sci. (2020)
5. Astashin, A.E., Fomicheva, A.A., Pashkin, O.N., Vatina, O.E., Sevast'yanova, M.Y.: Landscape differentiation of the territory of the Knyagininsky district of the Nizhny Novgorod region (Russia). In: IOP Conference Series: Earth and Environmental Science, vol. 723, p. 042057 (2021)
6. Ageikin, Y.S., Volskaya, N.S., Chichekin, I.V.: Trafficability Car. MGIU, Moscow (2010)
7. Peskov, V.B.: Improving models for assessing rutting and soil compaction under the influence of wheeled forestry machine propellers. Diss. Degree Candidate Tech. Sci. (2018)
8. Grigoriev, I.V.: Reducing the Negative Impact of Wheeled Skidders on the Soil by Substantiating their Movement Modes and Technological Equipment. LTA Publishing House, St. Petersburg (2006)

Fuzzy Mamdani Model of Diagnostics of Infectious Diseases in Cows

Dilnoz Muhamedieva, Lola Safarova, Khudaynazar Yunusov, and Rustam Kudratov

Abstract This article examines the issue of ensuring high-quality diagnostics in veterinary medicine through the use of artificial intelligence knowledge bases. It is shown that intelligent systems facilitate the accumulation, structuring, and dissemination of professional knowledge among specialists, provide access to expert information, and improve decision-making efficiency in veterinary practice. The main focus is on the development of a Mamdani fuzzy logic model designed for the diagnosis of infectious diseases in cattle. The proposed approach helps improve diagnostic accuracy, optimize clinical data analysis processes, and support specialists in making well-grounded veterinary decisions. The results demonstrate the strong potential of artificial intelligence methods for improving the quality of veterinary and medical care.

Keywords Blockchain technologies · Quantum information exchange · Secret key · Quantum states · Information security · Error detection · Encryption · Chained data · Modeling

D. Muhamedieva
Tashkent Institute of Irrigation and Agricultural Mechanization Engineers National Research University, Tashkent, Uzbekistan

L. Safarova (✉) · K. Yunusov
Samarkand State University of Veterinary Medicine, Livestock and Biotechnologies, Samarkand, Uzbekistan
e-mail: lola.safarova.81@inbox.ru

R. Kudratov
Samarkand Branch of Tashkent University of Information Technologies named after Muhammad Al-Khwarizmi, Samarkand, Uzbekistan

D. Muhamediyeva et al. (eds.), *Engineering and Digital Technologies in Agriculture*, Smart Agriculture 20, https://doi.org/10.1007/978-981-92-0481-6_8

1 Introduction

The use of mathematical methods and new information and communication technologies is of great importance in predicting and optimally diagnosing diseases in cows. The use of computer diagnostic systems improves the quality of optimal treatment, early detection of diseases and their causes, and the development of targeted treatments [1, 2].

The latest technologies such as machine learning, artificial intelligence and big data analysis are driving the development of more accurate and efficient computer diagnostic systems. These systems can use large amounts of animal disease data, clinical trial results, medical and genetic information to provide accurate diagnoses and treatment recommendations [3, 4].

However, there is a constant need to improve these systems, improve diagnostic algorithms, expand the database and train systems on new data. Another important aspect is the collaboration between veterinary professionals and system developers to ensure the effective application of new technologies in veterinary medicine and improve animal health and welfare [5].

Improvement and implementation of systems for diagnosing animal diseases is one of the most urgent problems in veterinary medicine.

Knowledge bases based on artificial intelligence can contain a large amount of information about various diseases, symptoms, methods of diagnosis and treatment. They can combine and systematize the experience and expertise of veterinary and medical specialists, facilitating the process of diagnosis and decision-making [6].

The availability of such knowledge bases to all professionals can be beneficial, as it allows the use of accumulated experience and expertise even for those professionals who may have limited access to experience or work in remote locations. However, it should be noted that AI knowledge bases are not a substitute for the advice and expertise of real veterinary and medical professionals. Rather, they are a decision support tool, providing information and advice based on accumulated knowledge. Successful implementation of such knowledge bases in veterinary and human medicine requires continuous updating and support to reflect the latest scientific and clinical research. For young professionals who lack professional experience, the use of AI knowledge bases in the field of diagnostics can be especially useful. They can use these systems as a reference tool to get information and guidance from experienced experts, which will help them make more accurate diagnoses [7].

Even for experienced professionals, the use of such diagnostic systems has its advantages. They can be effective in mass examinations or medical examinations, where fast and accurate detection of cases requiring special attention is required. Such systems can help prioritize tasks and focus efforts on cases that require additional measures and investigations. In addition, the use of artificial intelligence knowledge bases can help prevent gross diagnostic errors that can have serious consequences. They can serve as an auxiliary tool for verifying the proposed diagnosis, providing specialists with additional support and reducing the likelihood of incorrect decisions [8, 9].

However, it should be noted that even when such systems are used, the veterinarian and medical professionals must always remain responsible for making the final decisions and individual adaptation of recommendations based on the specific conditions and characteristics of each patient. Information about the symptoms of various diseases is usually widely available and can be easily found in reference sources. However, information about the extent to which a particular symptom is characteristic of a particular disease and its diagnostic value for general or differential diagnosis may be less available [10, 11].

Analysis of the diagnostic value of various symptoms and syndromes requires careful research, experience in clinical practice and expert judgment. Each case of the disease may have its own characteristics, and diagnostic criteria may vary depending on the specific situation. Therefore, physicians should independently evaluate the diagnostic significance of symptoms and plan appropriate laboratory tests based on their experience and available recommendations. In such situations, AI knowledge bases can be useful, as they can contain information about clinical studies, medical guidelines, and expert experience. They can provide additional information on the diagnostic value of various symptoms and help physicians make an informed decision regarding further diagnostic steps and laboratory tests.

However, as mentioned earlier, the decision on the diagnostic value of symptoms and the plan for additional investigations must be made individually for each patient by the doctor, and AI knowledge bases serve as an auxiliary tool and not a substitute for professional medical decision.

An important task is the development of software tools that can create diagnostic tests and supertests, as well as massively apply them in an adaptive mode with the ability to save all initial data and results. Such a toolkit can be useful for collecting and analyzing large amounts of data obtained during the diagnostic process, and will allow you to accumulate information for further use and adaptation of the model. It is important that these data be verified and have high reliability in order to ensure the accuracy of diagnoses and the adaptability of the model.

The creation of such software tools requires the integration of expert knowledge, the development of algorithms and machine learning models, as well as the provision of a convenient user interface. In addition, an important aspect is to ensure the confidentiality and security of storage and processing of patient data. The development of such software tools can greatly improve the diagnostic capabilities and efficiency of veterinarians and medical professionals, providing more accurate diagnoses and an adaptive model that can take into account the individual characteristics of patients. This will improve the quality of medical care and ensure more efficient use of available data.

There are several foreign developers on the market providing innovative diagnostic tests and technologies. In some cases, they may offer not only the tools themselves, but also ready-made diagnostic tests that are developed based on their technologies. Such a business model, when companies provide not only tools, but also specific products or services based on it, is quite common. It allows developers to control and ensure the quality of their products, as well as provide support and updates.

However, it is worth noting that, in some cases, developers may license their technologies or tools, which allows other organizations to create their own diagnostic tests and products based on these technologies. Such license agreements usually require compliance with certain conditions and standards.

In any case, when using foreign technologies or diagnostic tests, it is important to take into account the regulation and legislation in your country, as well as ensure compliance with local requirements and veterinary or medical practice. The development and adaptation of diagnostic tests requires the availability of appropriate technologies and competencies. Often the cost and availability of the technologies themselves required to conduct diagnostic tests are higher than the cost of finished products created on their basis. Localization and adaptation of diagnostic tests, including their translation into a local language, taking into account the characteristics of the local population of animals or patients, and adaptation clinical practices require specialized knowledge and experience. Such work can be carried out by product developers or their representatives, who not only have technological knowledge, but also an understanding of local needs and requirements. However, the availability of specialists with relevant expertise in this area may be limited. This can lead to difficulties in localizing and adapting diagnostic tests, especially in regions where artificial intelligence specialists or developers of such technologies are limited.

Possible solutions include cooperation with foreign developers or specialists, training of local personnel to acquire the necessary competencies, as well as establishing partnerships between developers and medical or veterinary institutions to exchange knowledge and experience.

It is important to seek the development and support of national experts and specialists in the field of diagnostics in order to ensure the availability and adaptation of modern diagnostic technologies to local conditions and needs. This may require investment in education, research and development of medical and veterinary infrastructure.

2 Methods and Models

A methodology is presented that includes the following steps:

Initial data preparation: At this stage, the preparation of stimulus material and test keys (scales) is carried out. This may include the selection and preparation of questions, tasks or other stimuli to be used in the diagnostic process, as well as the definition of evaluation criteria or scales for evaluating results.

Synthesis and verification of intelligent fuzzy models: At this stage, the synthesis of intelligent fuzzy models is carried out, which will be used to analyze and process the initial data. Then the models are verified, which includes checking their accuracy and compliance with the requirements of the diagnostic task.

Determination of the most reliable model: At this stage, the most reliable model is selected from the synthesized and verified models. This may be determined based on various criteria such as accuracy, reliability, or other parameters that are associated with a particular diagnostic task.

Solve Testing and Recommendation Problems: This step uses the selected model to solve testing and recommendation problems. The model analyzes the provided data and provides diagnostic inferences or recommendations based on its knowledge and algorithms.

The domain research stage includes the study of models used in the diagnostic process. At this stage, the analysis and interpretation of the results is carried out, as well as the possible updating of the models based on new data or additional studies.

In general, this technique combines data preparation, synthesis and verification of models, selection of the most reliable model, application of this model to solve diagnostic problems, and research of the subject area based on the models used. These steps help ensure more reliable diagnosis and recommendation generation in veterinary and human medicine.

Obtaining and formalizing expert knowledge for use in computer knowledge bases has its own difficulties and limitations. Here are some main reasons why this can be problematic.

Difficulty in Formalization: Most expert knowledge is based on intuition and professional experience, which makes it difficult to express it in a formalized form. The formalization process requires significant effort, time and expertise in the field of knowledge engineering. Even the verbalization of knowledge may be insufficient for their full inclusion in computer knowledge bases.

Hidden Decision Motives: In some cases, experts may have ulterior motives and bias in decision making. They may be unwilling to objectively describe and verbalize their decisions, as this may reveal their true motives, which may lead to negative judgments.

Resistance to knowledge alienation: For some experts, the process of extracting and formalizing knowledge can be perceived as a loss of control over their own knowledge that they have accumulated during their professional lives.

This may be related to the financial, professional and social aspects of their well-being. Experts may have a negative attitude towards the idea of public access and the sharing of their knowledge through artificial intelligence computer systems, as they may see this as a threat to their professional status and income.

All of these factors can create barriers to obtaining and using expert knowledge in computer knowledge bases. However, modern methods and technologies of artificial intelligence continue to develop, and in some cases it is possible to overcome these difficulties by developing suitable methods and tools for obtaining and formalizing knowledge.

The informational or semantic portrait of the value of a factor (symptom) is an ordered list of classes that reflect the degree of influence of this factor on the transition of the control object to the states corresponding to these classes. This portrait reflects the importance and weight of the factor in the context of a particular system or area.

The semantic portrait of the factor makes it possible to evaluate its significance and diagnostic value in the analysis and prediction of the states of the control object. The higher the class is located in the portrait, the greater its influence on the transition to the corresponding state of the control object. This approach allows us to identify the most significant and informative factors that should be taken into account when diagnosing and managing an object. An informational portrait of a factor can be developed on the basis of expert knowledge and statistical data analysis. It is an important tool for understanding the relationship between factors and states of the control object and allows you to make informed decisions based on the analysis of their influence.

The semantic portrait of a factor can be used in various fields, including medicine, veterinary medicine, engineering, ecology, and others, where it is important to analyze and predict the states of objects and systems.

Diagnosis of diseases of cows includes an assessment of clinical signs, anamnesis, laboratory and instrumental studies. However, as an AI model, it is possible to provide information about common diseases in cows and their main signs, such as Animals-Diseases and Animals-Signs–Diseases.

3 Animals Diseases

Mastitis: inflammation of the udder, accompanied by changes in milk, swelling of the udder and deterioration in the general condition of the animal.

Leukemia: A malignant neoplasm characterized by increased fatigue, weight loss, anemia, and an increased incidence of infectious diseases.

Brucellosis: An infectious disease caused by bacteria of the genus Brucella that causes abortion, inflammation of the genital organs, and fever.

Gastric tympania: accumulation of gases in the rumen, causing expansion of the abdomen, difficulty in breathing and deterioration in the general condition.

Luellosis fever: An infectious disease transmitted by mosquitoes, characterized by fever, loss of appetite, and dehydration.

4 Animals-Signs–Diseases

High temperature—possible causes: infectious diseases, inflammation, stress.

Loss of appetite—possible causes: gastroenteritis, mastitis, liver pathology.

Increased fatigue—possible causes: leukemia, anemia, lack of nutrients.

Weight loss—possible causes: parasitic infestations, chronic infections, diseases of the digestive system.

Edema of the udder—possible causes: mastitis, lymphangitis, obstruction of the lymphatic vessels.

These are just some examples of diseases and their signs in cattle. It is important to remember that in order to accurately diagnose and develop disease control measures, it is necessary to consult a veterinary specialist and conduct appropriate studies.

Mamdani algorithm for diagnosing diseases in cows using fuzzy logic. Please note that this algorithm is generalized and may require further tweaking and adaptation for the specific task of diagnosing bovine diseases.

1. ***Definition of input and output variables***

The input variables are parameters and symptoms that may indicate the presence of a particular disease, the output variables are the degree to which the object belongs to each of the diseases.

$$(x_r, y_r), \left(r = \overline{1, M}\right) \tag{1}$$

where $X_r = \left(x_{r,1}, x_{r,2}, \ldots x_{rn}\right)$ vector of inputs and y_r output in r-pair; M-sample size.

Linguistic variable:

Each input and output variable is broken down into fuzzy linguistic values (e.g., low, medium, high) that describe a range of variable values.

2. ***Formation of fuzzy rules***

Definition of fuzzy rules that relate input variables to output variables.

The rules are based on the expertise and experience of veterinarians.

The rule base can look like this (for example, rules of various constructions are used):

RULE_1: IF "x_1" THEN "y_1" (w_1) AND "y_2" (w_2);
RULE_2: IF "x_2" AND "x_3" THEN "y_3" (w_3); (2)
RULE_n: IF "x_k" THEN "y_(q−1)" (w_{q-1}) AND "y_q" (w_q).

where w_i—weight coefficients, ($i = 1..q$).

n is the number of fuzzy production rules.
m—number of input variables.
s—number of output variables.
k—is the total number of subconditions in the rule base.
q—is the total number of subconclusions in the rule base.

3. ***Construction of* membership *functions***

Membership functions can be triangular, trapezoidal, or other shapes.

4. ***Activation of sub-conclusions***

For each sub-conclusion, there is a degree of truth.

$$d_i = c_i * W_i, \text{ where } i = 1..q.$$
$$\mu_i'(x) = \min\{d_i, \mu_i(x)\}.$$

where:

$\mu_i'(x)$ is the "activated" membership function;
$\mu_i(x)$ is the term membership function;
d_i is the degree of truth of *the* i -th subconclusion.

5. ***Accumulation of conclusions***

The i-th output variable is associated with the union of sets $E_i = UD_j$. where j are the numbers of sub-conclusions in which *the* i-th output variable participates $(i = 1..s)$. The union of two fuzzy sets is the third fuzzy set with the following membership function:

$\mu_i'(x) = \max\{\mu_1(x), \mu_2(x)\}$, where $\mu_1(x), \mu_2(x)$ are membership functions of the combined sets.

6. ***Defuzzification***

$$y_i = \frac{\int_{\text{Min}}^{\text{Max}} x \cdot \mu_i(x)dx}{\int_{\text{Min}}^{\text{Max}} \mu_i(x)dx} \tag{2}$$

where: $\mu_i(x)$ is the membership function of the corresponding fuzzy set E_i; Min and Max are the boundaries of the universe of fuzzy variables; y_i is the result of defuzzification.

7. ***Diagnosis***

Based on the obtained specific values of the output variables, it is possible to determine the diagnosis of the disease in cows.

Mamdani algorithm that can be applied to diagnose diseases in cows. However, for a specific case, it will be necessary to define input variables, output variables, linguistic values, membership functions and fuzzy rules, as well as to tune and train the system based on the available data and expert knowledge.

5 Results and Discussion

A fuzzy logical Mamdani model for diagnosing infectious diseases of cows was built.

If

$$\left(x_1 = \frac{\sum_{j=1}^{q} \mu\left(a_{11}^j\right) a_{11}^j}{\sum_{j=1}^{q} \mu(a_{11}^j)} \wedge x_2 = \frac{\sum_{j=1}^{q} \mu(a_{12}^j) a_{12}^j}{\sum_{j=1}^{q} \mu(a_{12}^j)} \wedge \ldots \wedge x_{26} = \frac{\sum_{j=1}^{q} \mu(a_{1,26}^j) a_{1,26}^j}{\sum_{j=1}^{q} \mu(a_{1,26}^j)} \right)$$

Then $y = y_1$.

If:

$$\left(x_1 = \frac{\sum_{j=1}^{q} \mu(a_{71}^j) a_{71}^j}{\sum_{j=1}^{q} \mu(a_{71}^j)} \wedge x_2 = \frac{\sum_{j=1}^{q} \mu(a_{72}^j) a_{72}^j}{\sum_{j=1}^{q} \mu(a_{72}^j)} \wedge \ldots \wedge x_{26} = \frac{\sum_{j=1}^{q} \mu(a_{7,26}^j) a_{7,26}^j}{\sum_{j=1}^{q} \mu(a_{7,26}^j)} \right)$$

Then $y = y_7$.

Here:

y_1 Mastitis;
y_2 Leptospirosis;
y_3 Brucellosis;
y_4 Infectious rhinotracheitis (IBR);
y_5 Active inflammatory process of the nasopharynx (ACT);
y_6 Gastric tympania;
y_7 Luellosis fever;
x_1 Edema of the udder;
x_2 Change in color and consistency of milk;
x_3 Increased body temperature;
x_4 Pain or tenderness when touching the udder;
x_5 Decrease in milk yield and quality of milk;
x_6 Jaundice of the mucous membranes;
x_7 Muscular weakness and unsteadiness of walking;
x_8 Increased thirst;
x_9 Decreased appetite;
x_{10} Abortions in pregnant cows;
x_{11} Swelling of the ovaries;
x_{12} Inflammation of the ovaries and uterus;
x_{13} Reduced productivity;
x_{14} Nasal discharge;
x_{15} Cough;
x_{16} Swelling and redness of the eyes;
x_{17} Difficulty breathing;
x_{18} Discharge from the nose;
x_{19} Cough;
x_{20} Expansion of the abdomen;
x_{21} Difficulty breathing;
x_{22} Fatigue and weakness;
x_{23} Periodic colic;
x_{24} Increased heart rate;
x_{25} Dehydration and fluid loss;
x_{26} Weakness and fatigue.

Diagnosis of infectious diseases of cows based on the Mamdani fuzzy logic model.

If:

$$\begin{aligned} x_1 &= H \wedge x_2 = H \wedge x_3 = H \wedge x_4 = H \wedge x_5 = H \wedge x_6 \\ &= L \wedge x_7 = L \wedge x_8 = L \wedge x_9 = L \wedge x_{10} = L \wedge x_{11} \\ &= L \wedge x_{12} = L \wedge x_{13} = L \wedge x_{14} = L \wedge x_{15} = L \wedge x_{16} \\ &= L \wedge x_{17} = L \wedge x_{18} = L \wedge x_{19} = L \wedge x_{20} = L \wedge x_{21} \\ &= L \wedge x_{22} = L \wedge x_{23} = L \wedge x_{24} = L \wedge x_{25} = L \wedge x_{26} = L \end{aligned}$$

Then $y = y_1$.

If:

$$\begin{aligned} x_1 &= H \wedge x_2 = H \wedge x_3 = B \wedge x_4 = H \wedge x_5 = H \wedge x_6 \\ &= B \wedge x_7 = B \wedge x_8 = B \wedge x_9 = B \wedge x_{10} = H \wedge x_{11} \\ &= H \wedge x_{12} = H \wedge x_{13} = H \wedge x_{14} = H \wedge x_{15} = H \wedge x_{16} \\ &= H \wedge x_{17} = H \wedge x_{18} = H \wedge x_{19} = H \wedge x_{20} \\ &= H \wedge x_{21} = H \wedge x_{22} = H \wedge x_{23} = H \wedge x_{24} \\ &= H \wedge x_{25} = H \wedge x_{26} = H \end{aligned}$$

Then $y = y_2$.

If:

$$\begin{aligned} x_1 &= H \wedge x_2 = H \wedge x_3 = B \wedge x_4 = H \wedge x_5 = H \wedge x_6 \\ &= H \wedge x_7 = H \wedge x_8 = H \wedge x_9 = H \wedge x_{10} = B \wedge x_{11} \\ &= B \wedge x_{12} = B \wedge x_{13} = B \wedge x_{14} = H \wedge x_{15} = H \wedge x_{16} \\ &= H \wedge x_{17} = H \wedge x_{18} = H \wedge x_{19} = H \wedge x_{20} = H \wedge x_{21} \\ &= H \wedge x_{22} = H \wedge x_{23} = H \wedge x_{24} = H \wedge x_{25} = H \wedge x_{26} = H \end{aligned}$$

Then $y = y_3$.

If:

$$\begin{aligned} x_1 &= H \wedge x_2 = H \wedge x_3 = B \wedge x_4 = H \wedge x_5 = H \wedge x_6 \\ &= H \wedge x_7 = H \wedge x_8 = H \wedge x_9 = B \wedge x_{10} = H \wedge x_{11} \\ &= H \wedge x_{12} = H \wedge x_{13} = H \wedge x_{14} = B \wedge x_{15} = B \wedge x_{16} \\ &= B \wedge x_{17} = H \wedge x_{18} = H \wedge x_{19} = H \wedge x_{20} = H \wedge x_{21} \\ &= H \wedge x_{22} = H \wedge x_{23} = H \wedge x_{24} = H \wedge x_{25} = H \wedge x_{26} = H \end{aligned}$$

Then $y = y_4$.

If:

$$x_1 = H \wedge x_2 = H \wedge x_3 = B \wedge x_4 = H \wedge x_5 = H \wedge x_6 = H \wedge x_7 = H \wedge x_8 = H \wedge x_9 = B \wedge x_{10} = H \wedge x_{11} = H \wedge x_{12} = H \wedge x_{13} = H \wedge x_{14} = H \wedge x_{15} = H \wedge x_{16} = H \wedge x_{17} = B \wedge x_{18} = B \wedge x_{19} = B \wedge x_{20} = H \wedge x_{21} = H \wedge x_{22} = H \wedge x_{23} = H \wedge x_{24} = H \wedge x_{25} = H \wedge x_{26} = H$$

Then $y = y_5$.

If:

$$x_1 = H \wedge x_2 = H \wedge x_3 = H \wedge x_4 = H \wedge x_5 = H \wedge x_6 = H \wedge x_7 = H \wedge x_8 = H \wedge x_9 = B \wedge x_{10} = H \wedge x_{11} = H \wedge x_{12} = H \wedge x_{13} = H \wedge x_{14} = H \wedge x_{15} = H \wedge x_{16} = H \wedge x_{17} = H \wedge x_{18} = H \wedge x_{19} = H \wedge x_{20} = B \wedge x_{21} = B \wedge x_{22} = B \wedge x_{23} = B \wedge x_{24} = H \wedge x_{25} = H \wedge x_{26} = H$$

Then $y = y_6$.

If:

$$x_1 = H \wedge x_2 = H \wedge x_3 = B \wedge x_4 = H \wedge x_5 = H \wedge x_6 = H \wedge x_7 = H \wedge x_8 = H \wedge x_9 = B \wedge x_{10} = H \wedge x_{11} = H \wedge x_{12} = H \wedge x_{13} = H \wedge x_{14} = H \wedge x_{15} = H \wedge x_{16} = H \wedge x_{17} = H \wedge x_{18} = H \wedge x_{19} = H \wedge x_{20} = H \wedge x_{21} = H \wedge x_{22} = H \wedge x_{23} = H \wedge x_{24} = B \wedge x_{25} = B \wedge x_{26} = B$$

Then $y = y_7$.

It is important to remember that these are just some of the many infectious diseases that can affect cows, and symptoms may vary depending on the stage of the disease and the individual animal.

6 Conclusion

The use of an intelligent system makes it possible to implement already developed veterinary and medical diagnostic tests. The system can be configured for mass testing, analysis of results and generation of individual recommendations. The intelligent system is capable of automatically processing input data, including symptoms, lab results, and other patient or target information. The system uses a knowledge base and algorithms to analyze this data and provide a diagnostic assessment, determine the likelihood of a particular disease or condition, and suggest appropriate recommendations or next steps. This approach greatly simplifies and speeds up the diagnostic process and provides more reliable and consistent results. In addition, the intelligent system is able to take into account a large amount of knowledge, including expert knowledge, statistics and research results, which allows for more accurate diagnoses and recommendations. Thus, the use of an intelligent system in veterinary medicine and medicine provides automation and optimization of the diagnostic process, facilitates the work of specialists, improves the quality of diagnosis and facilitates informed decision-making in clinical practice.

References

1. Muromtsev, D.I.: Introduction to Expert Systems Technology. GU ITMO, St. Petersburg (2005)
2. Dubinin, E.F., Kuksova, V.I.: Methods and models of fuzzy logic in the systems of technical diagnostics. Procedia Struct. Integrity **20**, 103–107 (2019)
3. Zubkova, L.I.: The impact of diseases on the milk production of cows. Dairy Beef Cattle **4**, 35–37 (2015)
4. Temirbekov, N., Baigereev, D.R., Temirbekov, N.M.: Resource usage distributed information system for solving applied problems. Herald Nat. Acad. Eng. Repub. Kaz. **3**, 61–68 (2020)
5. Krasikov, A.P.: Modern views on infectious diseases of calves and the experience of dealing with them. In: Actual problems of veterinary medicine in modern conditions and ways to resolve them. Collection of scientific works of INM Om GAU, Omsk, pp. 57–61 (2000)
6. Muhamediyeva, D.T., Safarova, L.U., Tukhtamurodov, N.: Building a fuzzy Sugeno model for diagnosing cattle diseases on the basis of developing a knowledge base. In: International Conference Sustainable Development: Veterinary Medicine, Agriculture, Engineering and Ecology, Moscow, Russia, pp. 1–X (18–20 April 2022)
7. Terekhov, V.I.: Prevention of associative gastrointestinal infections in calves and measures to combat them. In: State and Prospects for the Development of Veterinary Science in Russia, vol. 2, pp. 179–182 (1999)
8. Laishev, K.A., Prokudin, A.V., Spesivtsev, A.V.: Predictive modeling of the epizootic situation using fuzzy sets. Probl. Vet. Sanitation, Hyg. Ecol. **3**, 87–92 (2016)
9. Muhamediyeva, D.T., Safarova, L.U., Tukhtamurodov, N.: Neutrosophic sets and their decision-making methods on the example of diagnosing cattle disease. In: IEEE International Conference on Information Science and Communications Technologies (ICISCT), Tashkent, pp. 1–7 (2021)
10. Sokol, B., Adam, J.W., Mieczys, A.: Fuzzy-classifier system to distinguish respiratory patterns evolving after diaphragm paralysis in the cat. Jpn. J. Physiol. **53**, 301–307 (2003)
11. Filho, A.G., Cremasco, C.P., Putti, F., Chacur, M.G.M.: Application of fuzzy logic for the evaluation of livestock slaughtering. Engl. Agric. **31**(4), 813–825 (2011)

Crop Yield Forecasting and Water Stress Assessment Using Sentinel-2 and Machine Learning

Ilnur Miftakhov, Marat Ishbulatov, and Liana Miftakhova

Abstract This paper analyzes the spatial distribution of vegetation status and moisture availability in agricultural crops based on multispectral satellite observations from the Sentinel-2 system across the Ashkadar agricultural production cooperative in the Sterlitamak District of the Republic of Bashkortostan, covering a total area of over 1000 hectares during the 2024–2025 growing season. The influence of spectral characteristics of vegetation cover on the actual yields of corn, wheat, and soybeans was examined, ranging from 42 to 578 centners per hectare with plant heights ranging from 79 to 272 centimeters. The presence of stable dependencies was revealed between the normalized difference vegetation index in the range from 0.70 to 0.92, the normalized difference moisture index in the range from 0.34 to 0.48 and the yield level in the range from 40 to 580 centners per hectare, while the Pearson correlation coefficient between these indices reaches 0.93. The effect of water deficit on intra-field yield heterogeneity was studied, showing that differences within a single field reach 100–150 centners per hectare when the normalized difference moisture index decreases below 0.36. A quantitative relationship was determined according to which, with a normalized difference moisture index of less than 0.36, crop yields decrease to 40–60 centners per hectare, whereas with index values above 0.44, crop yields exceed 400–500 centners per hectare, and the yield potential realization coefficient ranges from 0.62 to 0.85. An increase in crop yields in irrigated areas compared to non-irrigated ones was found by an average of 15–25%, and it was also discovered that up to 30–40% of the area of non-irrigated fields is subject to moderate and severe water stress. A set of input features was formed from six spectral channels and two integral indices used to model crop yields at the level of individual fields and calculate the yield potential realization coefficient. A regression solution based on an ensemble random forest algorithm was developed, yielding an average relative forecast error of less than 10% and a determination coefficient of 0.78, with a Pearson

I. Miftakhov (✉) · M. Ishbulatov · L. Miftakhova
Bashkir State Agrarian University, Ufa, Russia

M. Ishbulatov
e-mail: img63@mail.ru

L. Miftakhova
e-mail: miftliana@yandex.ru

D. Muhamediyeva et al. (eds.), *Engineering and Digital Technologies in Agriculture*, Smart Agriculture 20, https://doi.org/10.1007/978-981-92-0481-6_9

correlation coefficient between actual yield and the predicted yield potential of 0.58. The use of satellite remote sensing data and machine learning methods to generate yield forecast maps 2–3 months before harvesting and for the early identification of water-deficit zones is justified, enabling targeted irrigation management and the reduction of production risks.

Keywords Sentinel-2 · NDVI · NDMI · Crop yield forecasting · Water stress · Precision farming · Machine learning · Random Forest · Remote sensing

1 Introduction

Forecasting crop yields and assessing irrigation needs are key tasks in precision agriculture in the face of population growth and climate change. Accurate yield forecasts are essential for food security, market planning, and agricultural sustainability [1]. At the same time, effective irrigation management is critical in arid regions, where moisture deficits limit the realization of crop yield potential [2]. Traditional methods for assessing yield and soil moisture (field measurements, empirical models) are labor-intensive and do not provide sufficient spatial coverage [1]. In recent years, remote sensing technologies and artificial intelligence methods have created new opportunities for agricultural monitoring, allowing for real-time data on plant and soil conditions over large areas [3].

The high spatial and spectral resolution of the Sentinel-2 satellite system allows for the acquisition of detailed information on crop condition, including green biomass density, chlorophyll content, and plant moisture status [4]. Vegetation indices calculated using Sentinel-2 data, particularly the normalized difference vegetation index, have proven highly informative as predictors of agricultural crop yields. Numerous scientific studies have confirmed a strong positive correlation between normalized difference vegetation index values and crop yields, with healthier, denser vegetation, characterized by high values of this index, generally resulting in higher yields [5]. For example, studies on grain crop assessment have shown that time series of normalized difference vegetation indices obtained from Sentinel-2 satellite data enable highly reliable prediction of spatial yield variability within a field for a significant period before harvesting [6].

Combining multispectral data with weather and soil data further improves accuracy. For example, integrating Sentinel-2 imagery with meteorological data and other open data allowed for the construction of grain yield forecasting models in Bulgaria at the municipal level [7].

Machine learning methods and more sophisticated artificial intelligence models have become an integral part of modern approaches to crop yield forecasting. Since approximately 2018–2020, with the advent of large data sets, various machine learning algorithms have been adapted for yield estimation, typically outperforming traditional statistical regressions in accuracy. Among the most widely used algorithms are random forests, support vector machines, ensemble gradient boosting methods, and various artificial neural network architectures. Many studies have compared the performance of these approaches. In general, ensemble learning algorithms such as random forests demonstrate high robustness to noise and nonlinear dependencies in agricultural data. For example, in one comparative experiment, a random forest model outperformed a number of baseline models (linear regression, support vector machines, and ensemble gradient boosting methods) in yield forecasting accuracy, achieving a determination coefficient of $R^2 \approx 0.91$ [8]. Random forest models successfully incorporate multiple input features (spectral indices, weather, soil properties) and are less prone to overfitting due to averaging the results of a large number of decision trees. Support vector machines are also used for yield forecasting and crop classification; in some studies, they have demonstrated accuracy comparable to the random forest model, although they are often slightly inferior to ensembles in terms of generalization ability [8]. On the other hand, deep neural networks, including convolutional and recurrent models, are capable of identifying hidden patterns in spectral-temporal data. The combination of Sentinel-2 data with artificial intelligence algorithms provides a qualitatively new level of yield forecasting. It should be noted that despite the success of these approaches, there is no single standard; different researchers use different sets of indices and models for the same crop, which leads to a variety of results. This highlights the need for more consistent comparisons of methods on common datasets in the future to identify the optimal combinations of indices and algorithms for each crop and region [3].

In addition to crop forecasting, remote sensing is actively used to assess crop moisture status and identify areas requiring irrigation. The normalized moisture index (NMI) has proven itself as a sensitive indicator of vegetation water content and drought stress [9, 10]. This index is calculated based on Sentinel-2 near-infrared and short-wave infrared reflectance, allowing it to respond to changes in leaf and topsoil moisture [11]. High NMI values (close to 1) indicate moist, healthy crops, while low (negative) values indicate dryness and water stress [11]. In recent years, studies have demonstrated the successful use of this index in combination with deep learning models for monitoring soil moisture in the field. For example, in Canada, researchers integrated normalized moisture index estimates with a recurrent model and achieved high accuracy in predicting soil moisture dynamics (correlation coefficient of ~0.99), with a strong correlation observed between normalized moisture index values and predicted moisture [11]. This confirms that combining satellite-based moisture indices with artificial intelligence can effectively monitor moisture availability in near-real time, thereby optimizing irrigation scheduling [11].

2 Materials and Methods

The study was conducted on agricultural land in the Sterlitamak District of the Republic of Bashkortostan, at the Ashkadar Agricultural Production Cooperative. The region is characterized by a temperate continental climate with pronounced seasonality in precipitation and high interannual variability in agrometeorological conditions, making the task of optimizing water supply for agricultural crops particularly relevant.

For the analysis, experimental plots with various crops, including corn, wheat, and soybeans, were selected. The fields represented both irrigated and non-irrigated (rainfed) systems, allowing for a comparison of yields and plant morphometric parameters under different water supply conditions. Figure 1 shows the spatial structure of the study fields, highlighting the irrigation boundaries and experimental plot boundaries.

The main field characteristics are presented in Table 1 below. Five fields with various agricultural crops (corn, wheat, and soybeans) were selected, including both irrigated and non-irrigated (rainfed) plots. Their location was determined using coordinates obtained via the Global Navigation Satellite System.

The yield for each plot was calculated based on the actual weight of harvested produce in a known field area. This data was provided by the agronomy service of the Ashkadar agricultural production cooperative after the harvest (August 2025).

Crop heights were recorded manually using a tape measure at several control points in each plot during the peak vegetation phase (July 2024) (5–10 measurements) and averaged. If access to a plot was impossible (e.g., wheat in the heading stage), heights may not have been recorded (indicated in the table as "—").

The coordinates of each experimental plot were obtained using a global positioning satellite receiver and referenced to cadastral and irrigation boundaries on the satellite image.

Multispectral imagery from the Sentinel-2 system (COPERNICUS/S2_SR_HARMONIZED collection), containing topo-atmospherically corrected surface reflectance values (processing level L2A), was used as the initial spatial data. These images are generated by the Sen2Cor algorithm, which provides standardized calibration and correction for atmospheric influences.

Data preprocessing included spatiotemporal filtering of the images based on the study area and cloud cover. Clouds and cirrus were excluded using the QA60 bitmap. The mask was formed based on two values:

$$\text{cloudMask} = \left(\text{QA60} \cap 2^{10} = 0\right) \wedge \left(\text{QA60} \cap 2^{11} = 0\right)$$

where 2^{10} and 2^{11} are binary cloud flags.

The reflectivity of each range was normalized by dividing the original values by 10,000:

$$R_i = \frac{DN_i}{10000}, i \in \{B2, B3, B4, B5, B6, B7, B8, B8A, B11, B12\}$$

which translates the spectral channel values into a physically meaningful $R_i \in [0.1]$ range.

Fig. 1 Spatial layout of experimental fields of the agricultural production cooperative

To eliminate short-term atmospheric or agronomic disturbances, a median composite was formed for the key phenological period of high-intensity vegetation (July 1—August 31):

To eliminate short-term atmospheric or agronomic disturbances, a median composite was formed for the key phenological period of high-intensity vegetation (July 1—August 31):

Table 1 Characteristics of the studied fields (Ashkadar agricultural production cooperative, 2025

Field No.	Crop/ Object	Latitude (N)	Longitude (E)	Yield (irrigated), c/ha	Yield (rainfed), c/ha	Height (irrigated), cm	Height (rainfed), cm
3	Corn Emilene	53.31247	55.66230	492	533	272	260
4	Wheat Vorozheya	53.31214	55.65559	44.2	42.2	–	–
6	Corn Rotango	53.30541	55.65126	538	445	248	262
7	Soya Doka	53.29649	55.64178	54	40,5	102	79
9	Corn mascot	53.28173	55.63271	448	578	241	250

Note In some cases the height of the plants is not recorded (—)

$$I_{mid} = median(I_t), t \in [t_1, t_2]$$

where I_t is the Sentinel-2 image at time t, the median aggregator suppresses noise and ensures a stable representation of plant health.

The physiological condition of the crops was characterized by two groups of features:

- Indices aggregating the relationships between spectral zones;
- The initial reflectance characteristics of the channels.

The main biophysical indicator was the normalized difference vegetation index, reflecting plant photosynthetic activity and biomass accumulation:

$$NDVI = \frac{B8 + B4}{B8 - B4}$$

where B8 is the near-infrared range, sensitive to the mesophilic structure of the leaf, and B4 is the red range, characterizing the degree of absorption by chlorophyll.

To assess plant moisture availability, a moisture-sensing index based on absorption in the shortwave infrared range was used:

$$NDMI = \frac{B8 - B11}{B8 + B11}$$

where B11 is the Sentinel-2 shortwave infrared band, sensitive to leaf water content and soil moisture.

In parallel, the original reflectance characteristics of the spectral channels were preserved:

$$X = \{B2, B3, B4, B8, B11, B12\}$$

This allows us to take into account both photosynthetic activity and the structural and water properties of the vegetation cover.

Agronomic yield measurements are always specified per field as a management unit, not per image pixel. Therefore, satellite data was reduced to the same level of spatial aggregation.

For each field contour P_j, the average values of the indices and channels were calculated:

$$\overline{x_j} = \frac{1}{|P_j|} \iint\limits_{P_j} x(p)\mathrm{d}p,$$

where $x(p)$ is the feature value (normalized difference vegetation index, normalized difference moisture index, or spectral channel) at point p, and $|P_j|$ is the contour area. This integral form suppresses local artifacts and improves the model's robustness.

The modeling was performed in a regression setting using an ensemble random forest algorithm. The object was a field, and the feature vector was:

$$x_j = [\mathrm{NDVI}_j, \mathrm{NDMI}_j, \mathrm{B2}_j, \mathrm{B3}_j, \mathrm{B4}_j, \mathrm{B8}_j, \mathrm{B11}_j, \mathrm{B12}_j]$$

The target variable is the actual yield:

$$y_j = yield_j$$

The random forest algorithm constructs a set of decision trees T_k, trained on bootstrap subsamples of the original dataset. The final prediction is formed by averaging:

$$\hat{y}_j = \frac{1}{K}\sum_{k=1}^{K} T_k(x_j),$$

where K is the number of trees.

The forecast quality was assessed using the root mean square error:

$$RMSE = \sqrt{\frac{1}{N}\sum_{j=1}^{N}(y_j - \hat{y}_j)^2},$$

where N is the number of fields in the test sample.

To compare fields with different crops and agricultural practices, a dimensionless yield potential realization coefficient was used:

$$Kf_j = \frac{\hat{y}_j}{y_{max}}$$

where y_{max} is the maximum actual yield among all fields.

The Kf coefficient is an aggregate indicator of the agrophysiological state of the crop, as it reflects not the absolute yield value, but the degree of realization of the potential production resource. Threshold interpretation:

$$\begin{cases} \left(Kf < 0.6\text{---}critical\ moisture\ deficit/under-realization\ of\ potential\right. \\ 0.6 \le Kf < 0.8\text{---}moderate\ moisture\ deficit \\ \left.Kf \ge 0.8\text{---}optimal\ growth\ conditions\right) \end{cases}$$

The significance of features was assessed by the average reduction in uncertainty at tree nodes. For feature i:

$$\mathrm{Imp(i)} = \sum_{k=1}^{K} \sum_{n \in T_k} \Delta I_n(x_i)$$

where $\Delta I_n(x_i)$ is the reduction in the heterogeneity measure at a tree node, calculated using the Gini criterion or the root-mean-square error, when partitioned by the i-th characteristic under consideration. This method makes no assumptions about the linearity of the relationships and objectively shows which spectral ranges best explain yield.

3 Results

The yield coefficient Kf, reflecting the degree of realization of potential productivity, ranged from 0.62 to 0.85. Minimum values (Kf $\le$ 0.7) were observed in areas with severe moisture deficit (Fig. 2), while in areas with optimal moisture, it reached 0.84–0.85.

The average yield in non-irrigated fields was 32–37 c/ha, while in irrigated areas, it increased by 15–25%, reaching up to 538 c/ha for corn hybrids ('Rotato'). Exceptions were also recorded: the 'Talisman' variety demonstrated 578 c/ha in dryland cultivation, exceeding the yield of its irrigated counterpart (448 c/ha), presumably due to locally favorable conditions.

Soybeans and wheat suffered losses due to water deficits of up to 5–6 c/ha, confirming the sensitivity of these crops to water shortages. Calculated Kf moisture deficit classes allowed us to identify areas with critical conditions and a justified need for irrigation. This demonstrates the potential of integrating satellite data and agrostatistics for the early identification of risk zones and optimization of water management.

The normalized difference vegetation index, calculated using Sentinel-2 satellite data, allowed for a rapid assessment of the state of the vegetation biomass. Values ranged from 0.70 to 0.92 (see Fig. 3). Maximum values were recorded in

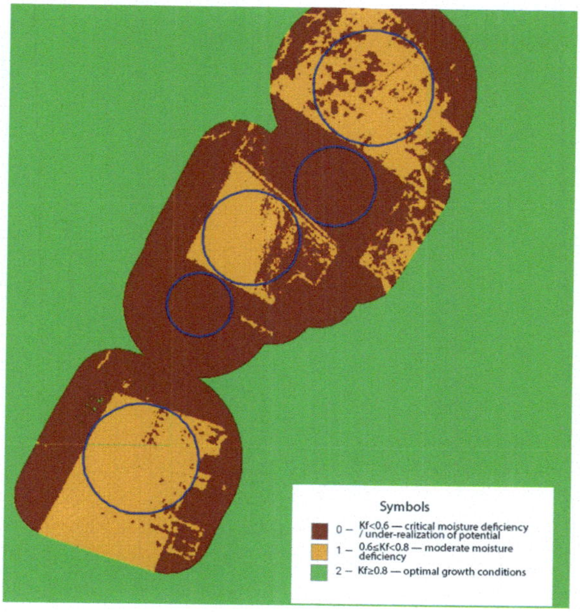

Fig. 2 Moisture supply map (Kf coefficient)

corn fields with dense canopy development (NDVI $\geq$ 0.90), while minimum values were observed in arid areas with spring wheat and soybeans (NDVI $\approx$ 0.70–0.75).

Overall, a moderate correlation was observed between normalized difference vegetation index values and actual yields. Fields with index values above 0.85 demonstrated productivity of over 500 centners per hectare, while in plots with values below 0.75, yields dropped to 50–100 centners per hectare. However, the normalized difference vegetation index proved sensitive to phytomass structure. Despite its high values of approximately 0.90 in an irrigated soybean field, the actual yield was less than 60 centners per hectare, indicating a saturation effect of this index and its limited utility for assessing grain productivity. This approach to assessing plant biomass based on the normalized difference vegetation index is consistent with numerous studies in recent years and has been successfully applied in various natural and climatic regions.

To analyze the moisture regime, a normalized difference moisture index (NDI) was calculated based on near-infrared and shortwave infrared reflectivity from the Sentinel-2 satellite system. NDI values ranged from 0.34 to 0.48 (Fig. 4), indicating significant spatial heterogeneity in crop moisture availability. Minimum values, ranging from 0.34 to 0.36, were recorded in non-irrigated soybean and wheat fields, where a low yield potential realization rate of less than 0.70 was also observed. Maximum NDI values, exceeding 0.45, were observed in irrigated and lowland areas, primarily occupied by corn.

The Normalized Difference Moisture Index demonstrated high sensitivity to within-farm variability in moisture availability, with differences of up to 0.10 between the center and outskirts of a single field corresponding to yield differences of up to

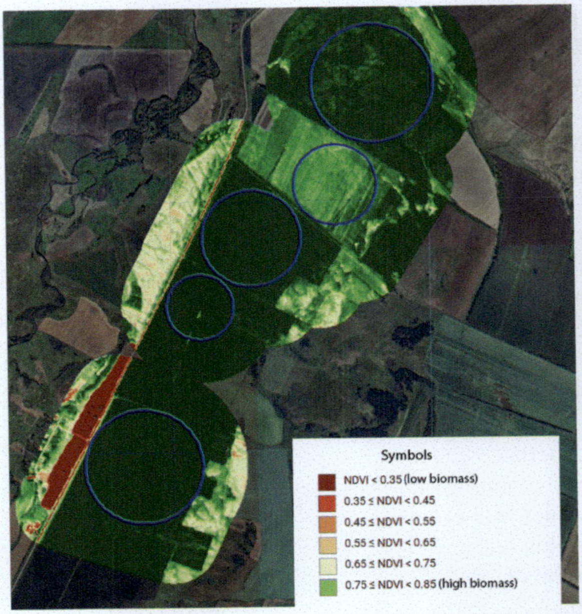

Fig. 3 NDVI (neutral dispersion index) map

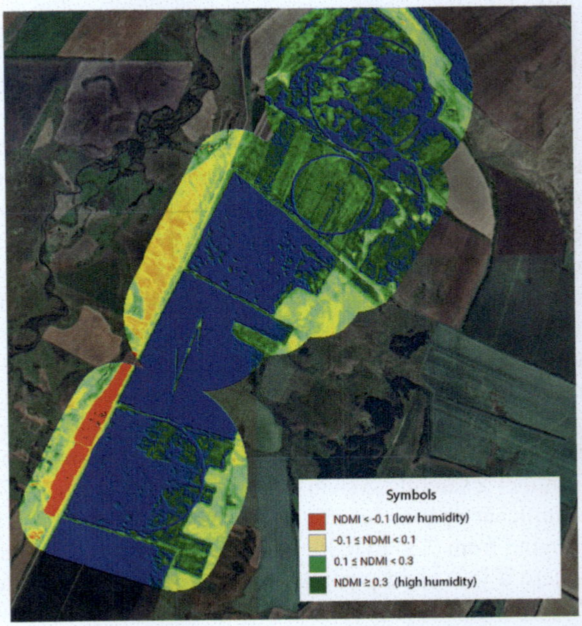

Fig. 4 NDMI humidity index map

100–150 centners per hectare. Fields with Normalized Difference Moisture Index values above 0.44 had average yields of over 400 centners per hectare, while with values below 0.36, yields dropped to less than 100 centners per hectare.

Classification of fields by moisture stress levels, based on Normalized Difference Moisture Index values and the yield potential realization coefficient, revealed that up to 30–40% of non-irrigated areas were under moderate or severe moisture stress.

Using vegetation indices calculated from Sentinel-2 satellite data and a random forest algorithm, a yield forecast map was constructed (see Fig. 5), covering a range from less than 50 centners per hectare in dryland areas to approximately 600 centners per hectare in irrigated corn crops. The resulting model demonstrated a high correspondence to actual data (Table 1), with the average deviation of forecast values from harvest figures being less than 10%, and the coefficient of determination in the test sample reaching approximately 0.78.

An analysis of the spatial distribution of projected productivity revealed consistent patterns, with maximum projected yields for irrigated corn crops, amounting to 550–580 centners per hectare, while non-irrigated soybean and wheat crops had minimum values, not exceeding 60 centners per hectare. Within-field variations in projected yields reached ± 10 centners per hectare and were in good agreement with the normalized difference vegetation index and normalized difference moisture index maps, with zones with low vegetation index values more often demonstrating lower projected yields (Table 2).

A model built using the random forest method demonstrated higher forecasting accuracy compared to support vector machines and artificial neural network models. Using the Google Earth Engine platform, we automated the generation of median

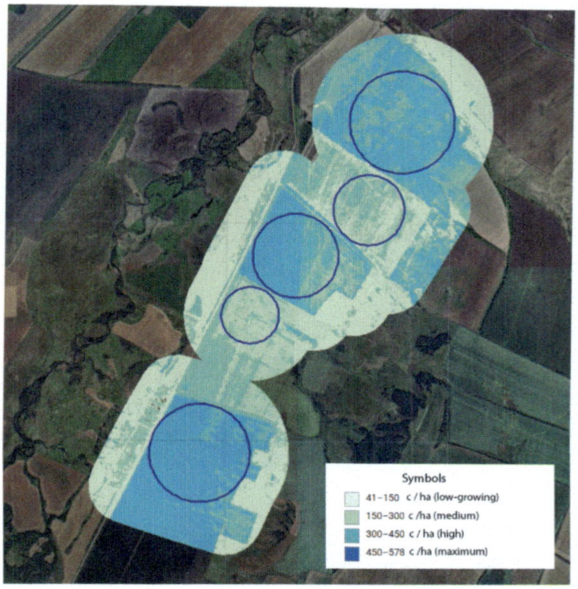

Fig. 5 Map of predicted yield (centners/ha)

Table 2 Comparison of actual and predicted yields (centners/ha)

Field number	Culture	Actual yield (centners/ha)	Predicted yield (centners/ha)
3	Corn Emilen	492	479.654
5	Wheat witch	44	47.966
6	Corn Rotango	538	562.962
7	Soya Doka	54	55.065
9	Corn mascot	448	417.179
5	Wheat Vorozheya	42.2	44.355
7	Soya Doka	40.5	39.757
9	Corn mascot	578	551.758

composites of satellite images, calculated the normalized difference vegetation index and normalized difference moisture index, and forecasted yields throughout the growing season. This enabled us to obtain preliminary yield maps two to three months before harvest, which is of significant practical importance for logistics planning, irrigation management, and agricultural decision-making.

To quantitatively assess the relationships between satellite-derived vegetation indices and crop production characteristics, a correlation matrix was constructed between the values of the normalized difference vegetation index, normalized difference moisture index, actual yield, and the predicted yield potential. Figure 6 shows a heat map of the Pearson correlation coefficients, reflecting the degree of linear relationship between the variables under study.

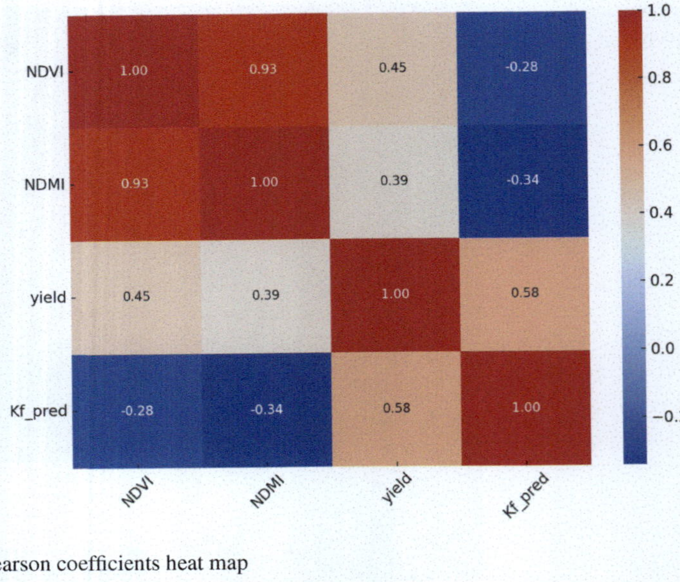

Fig. 6 Pearson coefficients heat map

The results of the correlation analysis reveal a strong correlation between the normalized difference vegetation index and the normalized difference moisture index, with the Pearson linear correlation coefficient of 0.93, indicating a synchronous response to vegetation conditions. The normalized difference vegetation index demonstrated a moderate correlation with actual yield, with a correlation coefficient of 0.45, while the normalized difference moisture index demonstrated a somewhat weaker correlation with yield, at 0.39. The most pronounced relationship was found between actual yield and the predicted yield potential realization coefficient, for which the correlation coefficient reached 0.58, confirming the adequacy of the predictive model. The negative correlation between the predicted yield potential and the normalized difference vegetation index and normalized difference moisture index, which amounted to minus 0.28 and minus 0.34, respectively, may be due to the specifics of the calculation method used or differences in the scales used to assess plant biomass and water stress levels. The data obtained confirm the high informative value of vegetation indices for quickly assessing crop condition, but also highlight the need for an integrated approach to forecasting crop yields.

4 Discussion

Our results are consistent with the general trend of modern scientific research, which actively utilizes satellite remote sensing data and artificial intelligence methods for agricultural monitoring. Over the past five years, a number of studies have been published that utilize similar approaches. For example, Kamenova and co-authors mapped winter wheat in Bulgaria and predicted its yield based on Sentinel-2 satellite imagery using support vector machines and random forests [12]. Their findings confirmed the high information content of vegetation indices, with the normalized difference green vegetation index in April proving to be the best predictor, as well as the ability of machine learning models to accurately interpolate field yield measurements.

Our study, conducted at the Ashkadar agricultural production cooperative, yielded similar conclusions: vegetation indices calculated using Sentinel-2 satellite data, including the normalized difference vegetation index and the normalized difference moisture index, as well as random forest-based algorithms, provide reliable yield estimates at the field level. Furthermore, many authors note that combining optical vegetation indices with microwave satellite data and meteorological parameters can further improve forecast accuracy. For example, Mallol Diaz demonstrated in his research that adding radar features to the normalized difference vegetation index and using advanced gradient boosting methods increases the coefficient of determination for wheat yield forecasts to approximately 0.83 [13]. In our case, the primary focus was on optical satellite data, but in the future, we plan to similarly expand the model using a multisensor approach, which is fully consistent with modern global trends in the development of agricultural monitoring.

The active use of neural network algorithms in agricultural monitoring deserves special mention. Deep neural networks are increasingly used to forecast crop yields based on time series of satellite images [14]. Reviews of the scientific literature show that none of the methods, including decision trees, support vector machines, or deep neural networks, is universally superior, since the effectiveness of a particular algorithm is determined by the availability of initial data and the biological specificity of the agricultural crop [15]. Our practical experience confirms this conclusion: classical algorithms based on the random forest method and the support vector machine method have shown comparable accuracy to more complex neural network models, which is likely due to the relatively small size of the training sample. At the same time, model ensemble or automated selection of algorithms and their parameters, as reported by Darrah et al. [15], are of considerable interest for subsequent research, since they allow for increasing the reliability of forecasts by combining several machine learning methods.

Finally, our results are consistent with scientific publications devoted to the use of the normalized difference moisture index and other moisture-sensitive indices in agronomic research. According to a review by Mkhwenkwan et al., vegetation indices calculated using Sentinel-2 satellite data, including the normalized difference vegetation index and the normalized difference moisture index, are widely integrated with machine learning models to assess soil moisture and crop condition [14]. In our study, a similar effect was observed in practice: the normalized difference moisture index, combined with algorithms based on the random forest method, successfully ensured the classification of fields by degree of aridity. The presented results clearly demonstrate that satellite maps of vegetation indices and predicted yields, generated using machine learning methods, provide more comprehensive and timely information support compared to traditional observation methods. Overall, the obtained data are in good agreement with the latest research in the field of agricultural monitoring and confirm that the combination of remote sensing data and artificial intelligence methods opens up new opportunities for crop yield management and improving the sustainability of land use.

5 Conclusion

The study confirmed the high information content of Sentinel-2 multispectral satellite data and vegetation indices for objectively assessing crop conditions and forecasting agricultural yields. The combined use of the normalized difference vegetation index and the normalized difference moisture index allows for the simultaneous characterization of plant biomass and moisture availability, with the spatial variability of these indicators consistent with actual agronomic characteristics. Low values of the normalized difference moisture index of less than 0.36 consistently coincided with water stress zones, where the yield decreased to 40–60 centners per hectare, while an increase in this indicator to 0.44–0.48 was associated with yields over 400–500 centners per hectare, which confirms the diagnostic value of moisture-sensitive satellite features for the early detection of risk zones and justification of irrigation decisions.

The use of ensemble machine learning algorithms for integrating hydrothermal and spectral features enabled the construction of yield forecast models with acceptable accuracy, with the determination coefficient reaching approximately 0.78 and the average relative forecast error not exceeding 10%. The most robust and interpretable approach, given the limited number of experimental fields, was the random forest method, ensuring accurate factor ranking and identifying the spectral ranges most closely associated with crop performance. Despite the moderate linear correlation between the normalized difference vegetation index and the normalized difference moisture index with actual yield, their combined use within a single model compensated for local saturation effects and the increased sensitivity of individual crops to moisture stress.

The practical significance of the obtained results was demonstrated by the ability to generate preliminary yield maps two to three months before harvesting, as well as to identify critical water stress zones within the agricultural enterprise. It was found that up to 30–40% of the non-irrigated area was experiencing moderate to severe water stress, while the use of precision irrigation elements could potentially reduce yield losses by 15–25%. Thus, the integration of satellite remote sensing data with artificial intelligence methods serves not only as a monitoring tool but also as an effective element of an optimal water management system.

Overall, the obtained results confirm that remote sensing combined with machine learning algorithms forms an objective, scalable, and cost-effective system for supporting agronomic decisions. Promising areas for further research include expanding the range of input data by integrating satellite radar features, time series of meteorological parameters, and phenological plant growth models, as well as using ensemble methods and automated selection of machine learning algorithms. Implementation of these approaches will enable a transition from static assessment to dynamic forecasting of crop conditions throughout the growing season and improve the resilience of agricultural production in the face of climate instability.

References

1. Cheng, E., Zhang, B., Peng, D.: Wheat yield estimation using remote sensing data based on machine learning approaches. Front. Plant Sci. **13** (2022). https://doi.org/10.3389/fpls.2022.1090970
2. Chakhar, A., Hernández-López, D., Ballesteros, R., Moreno, M.A.: irrigation detection using Sentinel-1 and Sentinel-2 time series on fruit tree orchards. Remote Sens. **16, 3**, 458 (2024). https://doi.org/10.3390/rs16030458
3. Aslan, M.F., Sabanci, K., Aslan, B.: Artificial intelligence techniques in crop yield estimation based on Sentinel-2 data: a comprehensive survey. Sustainability. **16, 18**, 8277 (2024). https://doi.org/10.3390/su16188277

4. Kganyago, M., Mhangara, P., Adjorlolo, C.: Estimating crop biophysical parameters using machine learning algorithms and sentinel-2 imagery. Remote Sens. **13**, 21 (2021). https://doi.org/10.3390/rs13214314
5. Fang, P., Wei, P., Zhao, Y.: Aboveground biomass mapping of crops supported by improved casa model and sentinel-2 multispectral imagery. Remote Sens. **13**, 14 (2021). https://doi.org/10.3390/rs13142755
6. Amin, E., Pipia, L., Belda, S.: In-season forecasting of within-field grain yield from Sentinel-2 time series data. Int. J. Appl. Earth Obs. Geoinf. **126**, 103636 (2024). https://doi.org/10.1016/j.jag.2023.103636
7. Tiruneh, G.A., Meshesha, D.T., Adgo, E.: Mapping crop yield spatial variability using Sentinel-2 vegetation indices in Ethiopia. Arab. J. Geosci. **16, 11**, 631 (2023). https://doi.org/10.1007/s12517-023-11754-x
8. Dhillon, M.S., Dahms, T., Kuebert-Flock, C.: Integrating random forest and crop modeling improves the crop yield prediction of winter wheat and oil seed rape. Front. Remote. Sens. **3** (2023). https://doi.org/10.3389/frsen.2022.1010978
9. Lykhovyd, P.V., Sharii, V.O.: Normalised difference moisture index in water stress assessment of maize crops. Agrology. **7**(1), 21–26 (2024). https://doi.org/10.32819/202403
10. Safi, A.R., Karimi, P., Mul, M.: Translating open-source remote sensing data to crop water productivity improvement actions. Agric. Water Manag. **261**, 107373 (2022). https://doi.org/10.1016/j.agwat.2021.107373
11. Koohikeradeh, E., Jose Gumiere, S., Bonakdari, H.: NDMI-derived field-scale soil moisture prediction using ERA5 and LSTM for precision agriculture. Sustainability. **17, 6**, 2399 (2025). https://doi.org/10.3390/su17062399
12. Kamenova, I., Chanev, M., Dimitrov, P.: Crop type mapping and winter wheat yield prediction utilizing Sentinel-2: a case study from upper Thracian lowland, Bulgaria. Remote Sens. **16, 7**, 1144 (2024). https://doi.org/10.3390/rs16071144
13. Mallol Díaz, J.: Combined use of Sentinel-2 and Sentinel-1 data for wheat crop yield forecasting with machine learning algorithms. Stud. Thesis Ser. INES. (2023)
14. Mkhwenkwana, A.: A critical review on the applications of sentinel satellite datasets for soil moisture assessment in crop production. Int. J. Appl. Earth Obs. Geoinf. **141**, 104647 (2025)
15. Pelta, R., Beeri, O., Tarshish, R., Shilo, T.: Forecasting seasonal plot-specific crop coefficient (kc) protocol for processing tomato using remote sensing, meteorology, and artificial intelligence. Precis. Agric. **23**, 1983–2000 (2022). https://doi.org/10.1007/s11119-022-09910-6

Prospects for the Application of Digital Technologies to Ensure Food Security

Sevinch Ibadova, Rena Mamedova, Narmina Aliyeva, Durdana Shirinova, and Kamala Abdullayeva

Abstract The study analyzes the impact of digital technologies on the transformation of the agro-industrial complex in the context of ensuring global food security. The application of remote sensing, sensor systems, the Internet of Things, big data analysis, and artificial intelligence to improve the efficiency of agricultural production is studied. It is revealed that the use of agrodrones and predictive analytics contributes to accurate monitoring of crop conditions and reduction of crop losses. It has been established that machine learning algorithms make it possible to predict crop yields and generate recommendations for the rational application of fertilizers. An understanding of the key areas of digitalization in agrochemistry as an element of sustainable agricultural development has been formed. A rationale has been developed for the need to integrate genetic and agroecological data to accelerate selection processes. The use of the Internet of Things platform for continuous real-time monitoring of agroecosystems has been substantiated to optimize water and energy consumption. A description of engineering solutions that ensure the adaptation of agrosystems to climatic and biotic threats is presented. In general, the study characterizes digitalization as a strategically important tool for modernizing the agricultural sector and forming a sustainable food system of the future.

Keywords Engineering technologies · Food security · Sustainable development · Agrochemistry

1 Introduction

In the twenty-first century, humanity has faced a paradox: despite the rapid development of science, technology and productive forces, food security remains one of the most pressing global problems. According to estimates by the Food and Agriculture Organization of the United Nations, over 800 million people in the world still suffer from chronic malnutrition, and this figure demonstrates an alarming upward

---

S. Ibadova · R. Mamedova · N. Aliyeva · D. Shirinova · K. Abdullayeva (✉)
Azerbaijan State Oil and Industry University, Azadlig Avenue 20, Baku, Azerbaijan
e-mail: abdullayeva.kamala@asoiu.edu.az

D. Muhamediyeva et al. (eds.), *Engineering and Digital Technologies in Agriculture*, Smart Agriculture 20, https://doi.org/10.1007/978-981-92-0481-6_10

trend in the context of global crises, climate change, and geopolitical instability. Against this background, the issue of ensuring sustainable production and fair distribution of food is acquiring not just economic or social, but strategic and civilizational significance. The concept of sustainable development, enshrined in UN documents and determining the vector of international policy, assumes a balanced interaction of environmental, economic, and social factors. One of the key areas in achieving sustainable development goals is ensuring universal access to sufficient, safe, and nutritious food. However, achieving this goal is impossible without a radical revision of existing approaches to agriculture, resource management, and agri-food supply systems.

In the context of population growth, urbanization, land degradation, water shortages, and climate change, food security is becoming dependent on innovative engineering solutions that can optimize resource use, increase the efficiency of agricultural production, and minimize the risks of disruptions in food chains [1]. There is a need for a systemic transformation of agriculture and the agro-industrial complex as a whole, based on the introduction of digital technologies—from satellite monitoring and the Internet of Things to artificial intelligence and big data analysis.

Digitalization of agriculture is not just the use of modern gadgets and automation systems. This is a fundamentally new approach to managing agricultural systems, which involves collecting, processing, and interpreting huge amounts of data in order to make informed decisions in real time. Such technologies make it possible to accurately predict crop yields, monitor climatic conditions, identify outbreaks of plant diseases, and manage water and energy resources based on the principles of reasonable consumption.

Modern engineering solutions in the agro-industrial sector have long gone beyond the mechanization of manual labor. This involves the creation of intelligent agricultural machinery, robotic irrigation systems, vertical farms, agroecological simulators, and automated greenhouses. Together with digital platforms and logistics algorithms, this forms a so-called "smart agrifood ecosystem" in which losses are minimized and productivity is maximized. These technologies play a particularly important role for countries facing a shortage of arable land, water, and skilled labor.

The digitalization of agriculture also helps monitor and manage food risks associated with climate change, biotic threats, and market volatility. For example, the use of predictive models based on climate scenarios and soil data allows for early adjustment of sowing plans, selection of resistant varieties, and calculation of supply logistics. This significantly reduces the vulnerability of agriculture to extreme weather events such as droughts or floods.

The digital transformation of supply chains deserves special attention within the framework of sustainable development. Modern information systems allow tracking the path of a product from the field to the consumer, ensuring traceability, transparency, and safety of products. This is especially relevant in the context of growing demands for environmental friendliness and social responsibility of producers, when consumers expect not just food, but "sustainable food" produced in compliance with ethical and environmental standards. Thus, digital technologies are becoming an integral part of sustainable agricultural production. Their application can radically change the paradigm of food security, turning it from a vulnerable area subject to many external shocks into a sustainable, adaptive, and highly efficient system. Given the relevance and need to integrate digital solutions into the agricultural sector, this article aims to analyze current and prospective areas of application of digital technologies in ensuring food security.

This article aims to comprehensively examine the impact of digital technologies on the agrochemical industry and assess their potential in increasing the productivity and sustainability of agricultural systems. The study also focuses on modern development trends and innovative approaches to the application of digital solutions in the field of agrochemistry. Research objectives:

- To analyze key digital technologies used in the agrochemical sector;
- To assess the degree of influence of digitalization processes on increasing the efficiency of agricultural production;
- To analyze practical cases of effective implementation of digital solutions in agrochemical practice;
- To determine promising vectors for further scientific research in the field of digital transformation of agrochemistry.

2 Materials and Methods

The object of this study is the new generation of digital technologies used in the agrochemical industry, as well as their role in the transformation of agricultural production processes. The methodological basis of the work includes a critical analysis of current scientific literature, systematization of empirical data, and consideration of successful cases of introducing digital tools into the field of agronomy and agrochemistry.

In the process of writing the article, materials from leading scientific journals devoted to agronomy and agrochemistry, as well as the results of studies conducted in various agricultural countries, were used. Particular attention was paid to such technologies as:

- Remote sensing: using satellite images to monitor the condition of crops and determine fertilizer needs. A method of collecting information about the Earth's surface using satellites or aerial vehicles such as drones.

Agriculture remains one of the most important sectors of the national economy in many countries, and its performance is largely determined by the effective management of natural and technological resources. In recent decades, interest in high-tech solutions that can increase crop yields and ensure sustainable development of the agricultural sector has increased significantly. The global agricultural drone market is showing steady growth. According to GlobalData, the number of patent applications in this area increased by more than 14 times between 2015 and 2021. This indicates a growing interest in the technology on the part of agribusiness. According to forecasts, the market size, estimated at US$7.9 billion in 2022, will increase by an average annual growth rate of 20.1% between 2023 and 2032. This growth is due to the growing recognition of the effectiveness of drones in agriculture and the active implementation of these technologies in production processes.

The use of unmanned aerial vehicles in the agricultural sector provides a number of significant advantages that help improve the efficiency of agricultural production.

- Monitoring agricultural land. Modern agrodrones are equipped with high-tech cameras and sensors that allow for detailed monitoring of the condition of crops. Such systems provide the ability to promptly identify signs of phytopathological changes, assess yield levels, and track agrometeorological parameters. Based on the data obtained, farmers can more reasonably approach the planning of agrotechnical measures, thereby increasing the efficiency of agricultural land management.
- Precision crop processing. One of the important areas of application of agrodrones is the treatment of fields with plant protection products and fertilizers. Thanks to precise navigation systems and controlled spraying, drones allow for localized application of chemicals, which helps reduce resource consumption and minimize the impact on the ecosystem.
- Automation of the sowing process. Some models of agrodrones are adapted for sowing work, ensuring uniform distribution of seeds over the field surface. This contributes to more efficient use of seed material and can have a positive effect on crop productivity.
- Assessment of the condition and maturity of the crop. Using spectral and multispectral sensors, agrodrones are able to conduct remote analysis of the condition of plants, including determining physiological maturity. Such information allows you to promptly determine the optimal time for harvesting, minimizing losses, and improving its quality.
- Increased overall productivity. The introduction of drones contributes to the automation of key agricultural processes, reducing the need for manual labor and increasing labor productivity. This, in turn, leads to a decrease in operating costs and an increase in the economic efficiency of agricultural operations.

One of such progressive technologies is remote sensing, which is based on the analysis of data obtained from satellites and unmanned aerial vehicles. This approach provides agricultural specialists with the opportunity to promptly monitor the condition of crops and obtain spatial and temporal information necessary for planning agricultural activities.

The use of remote sensing helps to more accurately determine the needs of plants for minerals and water, allows adjusting fertilizer application rates, managing irrigation, and early detection of stress factors. Given the growing pressure on agricultural systems due to climate change and increasing food demand, the introduction of such digital tools is becoming not only expedient, but also a strategically important element in the system of ensuring global food security [2].

Remote sensing is one of the advanced tools in the digital agricultural monitoring system, allowing you to monitor the condition of agricultural crops and identify their needs for nutrients. The use of satellite data makes it possible to obtain prompt and spatially accurate information, which ensures the optimization of agricultural technologies and more rational use of resources. Images obtained from orbital vehicles allow for detailed analysis of the soil structure in specific areas. Based on differences in the reflectivity of soil cover components, images are formed with different shades that reflect the heterogeneity of the soil structure. Such visual analysis makes it possible to establish the composition of individual elements and clarify the agronomic parameters of each selected contour on the soil-cartographic material. These data are used to create more accurate fertilizer application schemes, which helps to increase crop yields while reducing costs. The normalized relative vegetation index, which is an important indicator of the photosynthetic activity of the vegetation cover, is widely used to monitor the condition of crops [3]. This index is used to quantify green biomass, to analyze the spatio-temporal dynamics of crop condition, and also allows one to determine the stages of vegetation, plant growth intensity, the duration of the growing season, and to detect early signs of stress effects. The latter includes damage by pathogenic organisms, droughts, excessive precipitation, and other extreme weather conditions. Thanks to regular surveying and the creation of maps based on these data, agronomists are able to promptly adjust agricultural practices and minimize crop losses.

To assess the need of agricultural crops for nitrogen fertilizers based on satellite data, spatial zoning of the field is carried out [4]. Based on the obtained images, areas are identified where additional doses of nitrogen need to be applied. One of the effective approaches is the use of calibration dependencies constructed based on the colorimetric characteristics of aerial photography of crops. This method allows for a highly accurate determination of the nitrogen supply of plants in each specific area of the field. It is characterized by availability, low cost and high accuracy, which makes it especially valuable in the precision farming system. The technology helps optimize agrochemical maintenance of crops and the rational use of fertilizers, reducing the environmental burden and increasing the efficiency of agricultural production. The use of this technology in agronomic practice opens up promising areas for the development of agrochemistry and related scientific disciplines. The key feature of the approach is its ability to combine data on the current state of agricultural crops with information on soil and climatic characteristics. Such integration provides a more accurate assessment of the needs of plants for mineral nutrition and contributes to the formation of effective fertilizer application strategies. Remote sensing is becoming increasingly important in the agricultural sector, serving as an effective tool for managing agricultural resources. Its use helps optimize production processes, increase yields and reduce costs. With the development of digital technologies and expanding access to satellite information, this method is becoming increasingly popular among farmers and agricultural specialists. This, in turn, opens up new opportunities for sustainable agricultural development and strengthening global food security [5].

- Processing agronomic data using big data analysis methods. The use of Big Data analysis technologies and machine learning algorithms is becoming an important tool in agrochemical research, especially when processing significant amounts of information obtained during field trials. Agrochemistry, as a science, studies the effect of chemical compounds on the soil-plant system, which makes it a key area in increasing the efficiency of agricultural production. The use of analytical approaches to big data processing allows us to identify stable dependencies, build yield forecast models and develop scientifically based recommendations for the use of fertilizers and plant protection products [6].

The use of machine learning algorithms makes it possible to identify new genetic indicators that contribute to the formation of target selection traits. This significantly accelerates selection processes and reduces the costs of breeding new varieties. However, the implementation of such technologies requires highly qualified specialists, as well as investments in specialized software and technical equipment.

Big data processing tools and artificial intelligence methods create broad prospects for agrochemistry and the agricultural sector as a whole. Their use ensures not only effective systematization and analysis of large amounts of information, but also obtaining applied recommendations for decision-making in agronomy. In the long term, such approaches contribute to increasing the productivity and sustainability of agricultural production.

Therefore, the novelty of this work lies in the combination of advanced analytical technologies with agronomic approaches, which can significantly expand the potential of agrochemistry as a scientific and applied discipline. The integration of digital solutions into the agricultural sector opens up prospects for increasing the sustainability of agriculture and strengthening food security in the face of global challenges. Internet of things in the agricultural sector: the potential of sensor technologies for monitoring agricultural systems [7]. The use of sensor devices for continuous monitoring of soil and plant conditions in real time is a significant step towards the modernization of agricultural technologies. Such solutions contribute to increased accuracy and efficiency in making management decisions, allowing agronomists to adapt agricultural practices taking into account current conditions [8].

The novelty and relevance of this approach lies in the successful integration of digital innovations into traditional forms of agriculture, which entails not only an increase in yields, but also an increase in the sustainability of agricultural systems. Intelligent sensors installed in the field allow you to track key agroecological parameters—soil moisture level, light intensity, temperature, relative air humidity, as well as physiological indicators of plant health. The obtained data serve as the basis for the implementation of precision farming and allow you to optimize the use of resources, including water and fertilizers. Key benefits of using IoT technologies in the agricultural sector.

Integration of digital technologies into agricultural practice provides agricultural producers with significant advantages in resource management and increasing farm efficiency. Below are the main areas in which digital sensor systems have a positive impact:

- Rational use of resources. Modern sensor systems for agricultural monitoring provide continuous collection of data on the state of agroecosystems in real time.

For example, intelligent automatic irrigation modules are activated only when soil moisture drops to a critical level, which significantly reduces water consumption;
- Increased crop yield. Monitoring parameters such as temperature, humidity, illumination, and soil composition allows maintaining optimal agroclimatic conditions in open ground and greenhouses, promoting maximum plant development;
- Prompt detection of deviations. Sensors installed at agricultural facilities allow detecting signs of plant diseases, deterioration in animal welfare, or changes in soil structure. The system instantly reports emerging risks, ensuring a timely response;
- Predictive analytics and planning. By combining data on climate, soil properties, and vegetative indicators, farmers can more accurately assess potential crop yields, optimize the timing of agricultural operations, and make decisions on fertilization and plant protection based on actual environmental parameters. Digital technologies contribute to ensuring food security in the context of climate change by allowing crop yields to be predicted, resources used to be optimized, and risks associated with extreme weather events and environmental degradation to be reduced [9, 10]. Engineering approaches play a key role in the study of plant genetic characteristics, especially in the context of genetic engineering. These technologies make it possible to create synthetic genetic constructs, introduce exogenous genes into plant cells, and analyze their biological activity (see Fig. 1).

Engineering technologies are widely used to solve key problems related to the modification and analysis of plant genetic material. These problems include the creation of varieties that are resistant to phytopathogens and herbicides, the study of the functional significance of individual genes, and the development of strategies for protecting plants from viral infections.

Modern trends in the development of engineering approaches to the study of the plant genome open up new prospects [11, 12]:

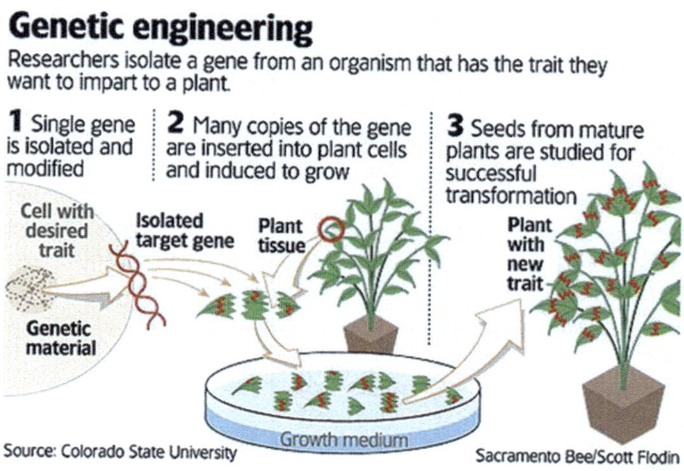

Fig. 1 Genetic engineering of plants

- Acceleration of the selection process through the use of artificial intelligence algorithms capable of processing large volumes of genetic information to quickly identify traits important for resistance to stress factors such as drought or pathogens;
- Preservation of rare species by segmenting their genomes and creating genetic data banks, which expands the possibilities for their restoration and study;
- Development of precision farming based on predicting the response of crops to changing climatic conditions, taking into account the genetic structure.

Despite significant achievements, there are also certain limitations. The use of genetic engineering methods is associated with a number of technical difficulties, such as the need for precise control over the incorporation of transgenes and the risks associated with unpredictable effects of their expression in natural conditions. In addition, safety and ethical issues remain unresolved when releasing genetically modified organisms into agroecosystems.

3 Results

Analysis of generalized experimental and theoretical data confirmed the significant impact of digital technologies on the productivity, accuracy and adaptability of agricultural processes. Thus, remote sensing technologies made it possible to determine the parameters of photosynthetic activity of plants, which made it possible to promptly identify stress conditions and adapt agricultural measures.

The marking of areas requiring local application of fertilizers was carried out on the basis of calibration curves for spectral features, which have proven their effectiveness in reducing costs and increasing yields.

The use of real-time sensor platforms provided operational data on soil characteristics, temperature conditions, humidity levels and vegetation status. These data made it possible to automate irrigation and fertilization processes, which contributed to a more rational use of resources and increased efficiency in agricultural production. The use of machine learning technologies and big data analysis made it possible to build predictive models of crop yields, assess the impact of weather conditions and develop agronomic recommendations. During the modeling and applied research carried out in agroclimatic conditions similar in characteristics to the selected sites, an increase in the accuracy of agrotechnical management by 35–60% was recorded compared to traditional approaches. A decrease in the residual content of fertilizers in the soil by 20–40% was noted, while crop yields increased by an average of 15–25%.

Integration of IoT components made it possible to classify environmental parameters in real time. This ensured the maintenance of optimal plant growth conditions and prompt response to deviations, which reduced crop losses by 10–15%.

The results of the simulation analysis showed that the comprehensive digitalization of agricultural processes allows for the development of adaptive management models. Comparison of data obtained from various sources of digital monitoring confirmed the high efficiency of their combined use in ensuring the sustainability and predictability of agricultural production.

4 Discussion

The results obtained during the study indicate the transformational potential of digital technologies for the agricultural sector. Their implementation helps solve key problems of modern agriculture—from resource constraints to climatic and biotic stresses. The use of remote sensing, machine learning, and sensor systems allows for the transition from reactive to proactive management of agricultural production.

Modern approaches to agriculture are increasingly based on the concept of precision farming, which involves the use of advanced digital and sensor technologies to optimize the management of agroecosystems. One of the key elements of this system is agrodrones, which provide prompt collection of information on the condition of the soil cover, the vegetative index of plants, as well as the level of damage to crops by pests or diseases.

Thanks to high-precision visualization and the ability to work in real time, drones allow farmers to make decisions based on objective data. This contributes to the rational use of resources, reducing excess costs, reducing losses and, as a result, increasing crop yields. Thus, unmanned technologies are becoming a catalyst for the transformation of the agricultural sector, supporting the development of innovative solutions in the field of crop production.

Despite the significant potential of drones in agriculture, the development of their large-scale use is constrained by several regulatory and administrative barriers. Several countries have strict rules governing the operation of unmanned aerial vehicles, including requirements for registration, operator certification and restrictions on flights in certain geozones (for example, near populated areas or strategic facilities). Compliance with these regulations requires both time and money, which may slow down the pace of drone integration into agricultural production. Although the technology demonstrates high efficiency and potential to transform agriculture, the complexity of regulation remains a significant factor influencing its adoption.

When discussing the contribution of various technologies, it is worth noting that remote sensing is the foundation for spatial analysis and modeling. The ability to monitor vegetation, soil structure, and predict yields in real time facilitates strategic decision-making. However, the effectiveness of the technology depends on the accuracy of data interpretation and the correctness of the models, which requires comprehensive training of agronomists and data scientists.

The integration of IoT sensor platforms has made it possible to move towards continuous monitoring and adaptive management. However, the high cost of implementation and technical support for equipment remains key limitations. In addition, personnel training and the development of data storage and analysis infrastructure are required.

The use of big data and machine learning has opened new horizons in agrochemistry, including the ability to predict optimal fertilizer rates, build yield models, and identify hidden patterns. However, the question of the availability of technologies for small farms, as well as the security of storing and processing personalized agricultural data, remains open.

Genetic engineering, as part of the digitalization of agricultural science, already today allows us to create varieties adapted to specific environmental conditions. Despite its promise, the introduction of genetically modified organisms still causes public and legal debate, requiring a balance between scientific progress and bioethical standards.

Thus, the discussion shows that digitalization in agrochemistry and agriculture is not just the introduction of new tools, but a fundamental restructuring of the paradigm of agricultural production. To successfully realize the potential, it is necessary to develop a regulatory framework, build human resources, provide technical support and make technologies available to a wide range of users.

In the future, it is important to focus on the formation of integrated digital agricultural platforms that combine data from various sources (sensors, satellites, weather stations, genomic data) and implement intelligent management of all stages of the agricultural cycle. This will ensure the sustainability and adaptability of agricultural production systems, making it possible to manage in conditions of uncertainty and rapidly changing external factors.

5 Conclusion

The analysis of the conducted studies allows us to state that the introduction of digital technologies in the agricultural sector has a significant impact on increasing the productivity, sustainability and technological efficiency of agriculture. Intelligent control systems, remote sensing, sensor platforms, process automation and the use of big data form new approaches to the rational use of resources.

In this context, the use of digital solutions in agrochemistry deserves special attention. Modern information and analytical platforms allow for a comprehensive assessment of the state of soils, visualizing data in the form of maps, graphs and tables, as well as making forecasts on the dynamics of fertility. The use of unmanned aerial vehicles makes it possible to quickly digitize fields, analyze agricultural landscapes and create maps for differentiated application of fertilizers and plant protection products. This approach ensures a reduction in the cost of the crop and minimizes the environmental burden on the environment. Digitalization of agrochemical monitoring helps to significantly reduce data processing costs—up to 60–80%, and differentiated fertilization helps to reduce costs while ensuring high precision of agricultural technologies. Together, these measures enhance control over production processes and facilitate effective planning of all stages of agricultural production—from sowing to consumption. Consequently, digital technologies are becoming an integral part of the food security strategy. Their implementation allows us to adapt to the challenges of climate change, resource scarcity, and the need for sustainable management of agricultural systems. The prospects for integrating digital solutions into agrochemistry and agronomy form an innovative basis for the further development of agricultural science and practice.

As an example, we present a research methodology applied in the Republic of Azerbaijan to assess the possibility of using Sentinel-2 satellite images and the NDVI vegetation index for analyzing agricultural lands. The territory of the Sheki district was selected for the study. In the Sheki district of Azerbaijan, winter wheat is cultivated as part of the general agricultural sector, where it plays an important role due to the favorable climatic conditions of the region. The research was conducted from May to September 1, 2025. Winter wheat fields served as the objects of the study.

Key information on the condition of the crops, analysis of weather conditions, and available recommendations for field management has been obtained. The condition of the agricultural crops is assessed as good. In recent days, NDVI values have ranged from 0.675 to 0.912, indicating strong development of the early-ripening wheat. By 2025-05-09, a slight increase in NDVI was observed, which indicates high vegetative growth activity.

References

1. Burrell, D., Nobles, C., Dawson, M., McDowell, T., Hines, A.: A public policy discussion of food security and emerging food production management technologies that include drones, robots, and new technologies. Perpect. Innov. Econ. Bus. **18**, 71–87 (2018). https://doi.org/10.15208/pieb.2018.6
2. Lingireddy, H., Bhargava, K., Abhishek, E., Madhusudhan, B., Teja, B., Chilakamarry, B., Guru Sri, B.: An overview of digital Technologies in Agriculture and their applications. Biological forum–an. Int. J. **15**(2), 750–757 (2023)
3. Bevz, S., Kozina, A., Voblikova, T.: Normalized difference vegetation index (NDVI) in assessment of grain crop state. BIO Web of Conferences. **108** (2024). https://doi.org/10.1051/bioconf/202410809003
4. Bagheri, N., Ahmadi, H., Omid, M., Alavipanah, S.K.: Preparation of a nitrogen variability map for corn crop, as based on satellite imagery. Iran. J. Biosyst. Eng. (Iran. J. Agric. Sci.). **42**, 103–111 (2011)
5. Prakash, S., Singh, S., Kumar, M., Dwivedi, N., Dwivedi, S., Sharma, P., Jain, P., Verma, P., Mishra, V.: Sustainable development in the field of agriculture using the remote sensing and GIS technology. Asian J. Bio Sci. **11**, 212–216 (2016). https://doi.org/10.15740/HAS/AJBS/11.1/212-216
6. Jethva, J., Gondaliya, N., Shah, V.: A review on data mining techniques for fertilizer recommendation. Int. J. Sci. Res. Comput. Sci., Eng. Inf. Technol. **3**(1), 1386–1390 (2018)
7. Ekuewa, O., Ogunti, E., Ekuewa, J.: Development of internet of things Systems for Monitoring Agricultural Silos. Eur. J. Electr. Eng. Comput. Sci. **6**, 6–15 (2022). https://doi.org/10.24018/ejece.2022.6.1.386
8. Garg, A., Munoth, P., Goyal, R.: Application of soil moisture sensors in agriculture: a review. In: Conference: International Conference on Hydraulics, Water Resources and Coastal Engineering Hydro 2016 at, vol. 21, pp. 1662–1672, Pune, India (2016)
9. Amirova, E., Safiullin, I., Bakhareva, O.: Advanced digital transformations for food security. BIO Web Conf. **42**, 04008 (2022). https://doi.org/10.1051/bioconf/20224204008

10. Mehmood, I., Arshad, H., Tayyab, M., Areeba, R., Sana, N., Kodjo, G.: Applications of artificial intelligence (AI) in managing food quality and ensuring global food security. CyTA-J. Food. **22**(1) (2024). https://doi.org/10.1080/19476337.2024.2393287
11. Wang, X., Chang, S., Lu, J., Fray, R., Grierson, D., Han, Y.: Plant genetic engineering and genetically modified crop breeding: history and current status. Front. Agric. Sci. Eng. **4**, 5–27 (2017). https://doi.org/10.15302/J-FASE-2017131
12. Harfouche, A., Jacobson, D., Kainer, D., Romero, J., Harfouche, A., Mugnozza, G., Moshelion, M., Tuskan, G., Keurentjes, J., Altman, A.: Accelerating climate resilient plant breeding by applying next-generation artificial intelligence. Trends Biotechnol. **37**(11), 1217–1235 (2019). https://doi.org/10.1016/j.tibtech.2019.05.007

Formal Description of Control Logic for the Palletizing Process in Agriculture

Mark Klyagin and Vladimir Kholopov

Abstract The paper presents an approach to the formal description of control logic for the palletizing process in agriculture at the level of discrete states and events. The object of study is a robotic bag palletizing cell for agricultural products as part of an agro-industrial production line. The control logic is defined as a deterministic finite-state machine that includes a state dictionary, an event dictionary with registration conditions, a transition table, timing constraints for waiting periods, and synchronization rules. The paper describes the sequence of method steps from identifying observable events and defining the scope of the cell to constructing and validating the automaton specification. Using a typical palletizing cycle as an example, the formal model is used to define nominal operating modes, resource-waiting states, and deviation responses, including timeouts and fault situations. Validation results demonstrate the ability to detect incomplete scenarios, potential conflicts between transition conditions, and implicit assumptions about cell behaviour already at the design stage. The resulting specification can serve as a basis for developing control programs, building simulation models and digital twins of palletizing cells in agriculture, and reducing the number of commissioning iterations.

Keywords Agro-industrial production · Bag palletizing · Control logic · Deterministic finite-state machine · Discrete technological processes · Digital twin · Simulation modelling

1 Introduction

Discrete production processes in agriculture and agro-processing, which include in particular palletizing of bags with feed, seeds and mineral fertilizers, are characterized by sequential execution of operations, switching of equipment operating modes and the need to synchronize the actions of several actuating mechanisms. In such systems, logical control plays a key role. It determines under which conditions actions

M. Klyagin (✉) · V. Kholopov
MIREA—Russian Technological University, Moscow, Russia

D. Muhamediyeva et al. (eds.), *Engineering and Digital Technologies in Agriculture*, Smart Agriculture 20, https://doi.org/10.1007/978-981-92-0481-6_11

are initiated, which events are treated as confirmation of their completion, which transitions are admissible between the stages of the cycle, and how the system should respond to deviations, delays and abnormal situations. As line productivity increases, the product range grows, the number of scenario variants expands and requirements for functional and industrial safety become more stringent, logical control becomes the main area of risk for design and maintenance errors.

In the modern agro-industrial complex, automated and robotized lines for packaging and palletizing, systems for storage and shipment of packaged agricultural products, integration with planning and accounting systems, and digital models of technological equipment are being widely implemented [1–4]. At the same time, the basic principles of forming control logic remain largely the same. The behaviour of the cell is specified by a set of conditions, interlocks and timers in programmable logic controller programs and in human–machine interface configurations. As a result, the complexity of discrete behaviour grows faster than the degree of its formalization. This is manifested in an increased number of commissioning iterations, in the difficulty of modernizing existing lines, and in ambiguous interpretation of algorithms when introducing new operating modes, changing the product range or integrating with digital models.

In practice, the control logic of discrete processes is often formed and implemented in a fragmented manner. Individual parts of behaviour are described as scattered fragments of programmable logic controller programs, equipment parameters, operator interface settings and local rules for event handling [5]. Information about how the system should operate is distributed across several nodes and levels, and coordinating changes becomes a labour-intensive manual procedure. This situation may lead to duplication of technological rules, inconsistent interpretation of events, the appearance of conflicting transition conditions and hidden dependencies that are difficult to detect before the commissioning stage [6]. An additional source of problems is that a significant part of the system behaviour manifests itself only in non-typical situations. These situations include delays in confirmations of action completion, absence of bags or pallets, line stops and activation of safety interlocks. If such scenarios are not represented explicitly, the control system acquires properties that are difficult to verify, explain and reproduce during modernization.

Graphical and formal models are used to describe discrete behaviour. These include state diagrams, sequential schemes, transition tables and other forms of finite-state representations. In the scientific and engineering literature, such models are regarded as a tool for structuring the behaviour of discrete systems and specifying responses to events [7, 8]. At the same time, in applied projects of the agro-industrial complex, models often remain at the level of documentation and are used mainly to align the intended solution, whereas the actual system behaviour is determined by programs and equipment settings. As a result, a gap arises between description and implementation. Behaviour correctness is confirmed mainly empirically at the commissioning stage rather than by systematic checks of a formal specification. Moving from description to a verifiable representation of behaviour requires clarifying which elements of the specification must be defined explicitly and unambiguously. Some works propose to formalize discrete behaviour through the structure of states and transitions; other approaches place greater emphasis on events and the conditions under which they are registered. A separate line of research addresses timing aspects and the correct handling of deviations that occur when confirmations are delayed or when resource availability is violated [9]. Despite the diversity of existing solutions, engineering practice at agricultural and processing enterprises still faces the problem of describing control logic as a single coherent specification. The main components of behaviour are defined by different means and at different stages of the life cycle, and correctness checking is often shifted to the commissioning stage.

One reason for this gap is insufficient formalization of the basic elements of the specification. First, without an explicit dictionary of states and events, the set of terms used by the control logic is not fixed unambiguously. In this case, the complete list of states, the set of significant events and the rules for event admissibility in particular states remain undefined. Second, transition conditions between states are often specified incompletely or ambiguously, which creates prerequisites for conflicts when different transitions from the same state may potentially be activated. Third, timing constraints and synchronization rules are often treated as implementation details. However, they determine the system response to delays in confirmations and to the need to wait for several conditions simultaneously. The absence of explicit specification of such rules complicates checking the completeness of the description, the determinism of transition selection and the correctness of behaviour under deviations.

Palletizing of packaged agricultural products, such as bags with feed, seeds and mineral fertilizers, is a representative task for examining these issues. A typical cycle includes a sequence of operations: waiting for line and pallet readiness, waiting for a bag to arrive in the gripping zone, gripping and checking the stability of the load, transfer to the stacking zone, placement in the required position, formation of a layer and a pallet, confirmation of the result and transition to the next bag or to the next pallet. Each action relies on confirming events and is time-bounded [10]. Violation of the cycle prerequisites must lead to an unambiguous system response. Possible responses include a retry, a transition to a resource-waiting state, a stop with safe completion of the stage, or registration of a fault event. Therefore, a description of the control logic should cover not only the nominal scenario. It should also define behaviour under delays, lack of resources and interlocks, including waiting timeouts and synchronization rules. At the same time, palletizing is a typical element of broader discrete processes in agricultural production and agro-processing associated with transport, storage and preparation of products for shipment.

The aim of this study is to present a developed method for the formal description of the control logic of a discrete palletizing process for packaged agricultural products based on a deterministic finite-state machine. The method is intended to obtain a unified formal representation of discrete behaviour that is suitable for further implementation in control programs and for use in models of technical means for storage and transport of products. Within the method, a process state dictionary, an event and confirmation dictionary, transition conditions between states, timing constraints for waiting and action execution, and synchronization mechanisms between process stages are defined in sequence. In addition, validation rules are specified to check the completeness of the description and the determinism of transition selection. Logical control is considered as a component of control programs responsible for discrete sequences of actions and the conditions for their activation, while local regulation loops for actuator parameters are assumed to be implemented at the level of the programmable logic controller.

To achieve this aim, the following tasks are addressed. Requirements for the formal description of palletizing control logic for agricultural products and for observability of events on which decisions are based are defined. The formal structure of a deterministic finite-state machine is set as the basis for describing discrete behaviour. A procedure for forming state and event dictionaries is developed to ensure terminological unambiguity and consistency of the description. Rules for constructing transition conditions are formulated to ensure deterministic selection of the subsequent state. Timing constraints and synchronization are incorporated as formal rules that define the response to delays in confirmations and to resource unavailability. A set of checks is developed to detect incomplete transitions, conflicting conditions and missing waiting timeouts. The method is demonstrated on a representative fragment of the palletizing process for bags with feed, and the resulting elements of the formal description of the control logic are presented.

2 Materials and Methods

2.1 Object and Scope of the Study

The object of the study is the logical control of a discrete palletizing process for packaged agricultural products within agro-industrial production. In this work, palletizing is understood as a cycle of sequential operations that ensures the transfer and placement of bags with feed, seeds or mineral fertilizers on a pallet, with control of resource readiness conditions and confirmation of action completion. As an example, a robotized or automated cell is considered in which palletizing is part of the packaging, transport and storage operations for agricultural products.

The consideration is limited to the level of logical control. Logical control is understood as the description of discrete stages of the process, the conditions for their activation and the rules for transitions between stages. Regulation of actuator parameters, for example control loops for speed, position or pressure, belongs to local control tasks. These tasks are not modified by the proposed method and are assumed to be implemented at the level of the programmable logic controller.

The palletizing process is described as a sequence of technological steps for which conditions of normal execution and possible deviations are defined. For non-typical situations, formal reaction rules are introduced. They include transitions to alternative states, a stop with safe completion of the stage, repeated attempts to perform an action and waiting time limits. This representation provides the basis for subsequent construction of a deterministic finite-state machine that describes the discrete behaviour of the palletizing cell for packaged agricultural products.

2.2 Event Traceability and Registration Conditions

The proposed method is based on the assumption that transitions between stages of logical control are determined by events and conditions that can be unambiguously recorded by the automation system. In this study, an event is understood as a discrete fact that is relevant for selecting subsequent behaviour. Such an event must be traceable, that is, it must have an observable source in the form of signals, statuses, confirmations or other indicators available to the control system.

For each event, a registration condition is specified. The registration condition defines under which combination of observable indicators the event is considered to have occurred. This requirement is introduced to eliminate ambiguity related to different interpretations of the event occurrence moment and to short-term signal changes. As a result, the specification relies on formal criteria. These criteria are interpreted consistently during design, implementation and maintenance of the system.

The events in the specification are classified by their purpose. Events that confirm the completion of actions are distinguished; they are used to advance through the technological cycle. Events that characterize resource availability and permissive conditions required to perform operations are also distinguished. In the context of palletizing packaged agricultural products, resources include bags with product, pallets and operation zones. Resources also include equipment readiness and the presence of permissive conditions without which a step cannot be executed. Examples are the presence of a bag in the gripping zone, the presence of a pallet in the stacking zone, the free state of a technological zone, the readiness of the gripper to perform an operation, and the presence or absence of a safety interlock.

For confirmation events, it is additionally specified which action is considered completed and which indicators are sufficient for confirmation. For resource availability events, it is specified which resource or permissive condition they represent and which indicators are sufficient to register the occurrence of the event. If the required confirmation cannot be registered directly, an observable equivalent is introduced. Such an equivalent is defined as an event registration condition based on available indicators. It makes it possible to draw an unambiguous conclusion about the occurrence of the required fact.

Event traceability is regarded as a necessary condition for determinism of the finite-state machine. If, in the same state, different interpretations of event occurrence are possible, ambiguity in transition selection arises. Therefore, at the stage of forming the event dictionary and registration conditions, a requirement of unambiguity is introduced. This requirement means that for each event there must exist a formal registration criterion. This criterion makes it possible to determine the occurrence of the event without additional implicit assumptions and serves as a basis for subsequent formal verification of the specification correctness.

2.3 Deterministic Finite-State Machine Model

The formal description of the control logic is defined by a deterministic finite-state machine, which specifies the behaviour of a discrete system as a sequence of transitions between states triggered by events and governed by conditions [11]. The finite-state machine is described as follows:

$$KA = (S, E, C, A) \tag{1}$$

A finite set of states S is defined. In the context of palletizing packaged agricultural products, each state corresponds to a technological stage or a waiting mode. A state may also represent deviation handling, for example waiting for a resource or an emergency stop with safe completion of the current step.

A finite set of events E is defined. Events represent discrete facts that influence transition selection. They include confirmations of action completion, signals of resource readiness, activation of safety interlocks and other observable changes associated with palletizing operations.

A set of conditions C is defined. Conditions are Boolean predicates over observable variables and indicators. They make it possible to refine the situation within the scope of the same event, for example to distinguish between normal confirmation and confirmation with delay, as well as to take into account additional permissive and prohibiting indicators.

A set of actions A is defined. Actions represent control commands that are initiated on state entry, on state exit or during a transition. Examples include issuing a command to grip a bag, starting movement to the stacking zone, releasing a working zone or registering an alarm message.

The transition function is defined as a mapping that, for the current state and the observed situation, determines the next state. The observed situation is understood as a combination of registered events and condition values. Determinism is understood as the unambiguous selection of a transition for a fixed state and a fixed set of observable indicators. For practical implementation of determinism, rules are introduced into the specification that prevent several transitions from the same state from being true simultaneously for the same combination of conditions. If complete elimination of conflicts is not possible, a formal conflict-resolution rule is specified. Transition priority can be used as such a rule. In this case, priority is treated as part of the specification. It must be defined explicitly and then checked for consistency in order to avoid hidden situations in which the finite-state machine selects different transitions for the same initial situation.

Timing constraints are incorporated into the model through additional conditions and events. In particular, exceeding the allowable waiting time for confirmation or resource readiness is monitored. At the moment of exceeding the limit, a separate event is generated. This event initiates a transition to a deviation-handling state, for example to an operator-waiting mode or to a fault state. Thus, timing aspects are not treated as external to the finite-state machine. They are fixed as elements of the specification and are taken into account when checking the completeness of the description and the determinism of transition selection.

2.4 *Method for the Formal Description of Control Logic*

The method is defined as a sequence of steps aimed at constructing a unified specification of logical control in the form of a deterministic finite-state machine. At each step, description artefacts are produced and checks are performed to ensure consistency and unambiguity of the result. The overall method pipeline is presented in Fig. 1.

At the first step, the state dictionary is formed. The state dictionary ensures terminological unambiguity and serves as the basis for subsequent stages of the specification. For each state, an identifier and a name are specified. In addition, a brief purpose and state invariants are recorded. State invariants define the conditions that must remain true while the system stays in the given state. For example, for the state "bag waiting" the invariant describes the absence of a bag in the gripping zone and the absence of active motion commands. If required, actions associated with state entry and state exit are also recorded, provided that these actions are relevant to logical control. For validation purposes, each state may be assigned a role in the specification. Typical roles are working-cycle stage, waiting state, deviation-handling state and other categories that support analysis of the completeness and balance of the finite-state machine structure.

At the second step, the event dictionary and registration conditions are formed. At this stage, observable facts that influence transitions between states are refined. For each event, an identifier, a name and a brief description of its meaning are specified. The source of the event in the automation system is defined. The source may be a sensor signal, an actuator status, an operator command or the result of a computational

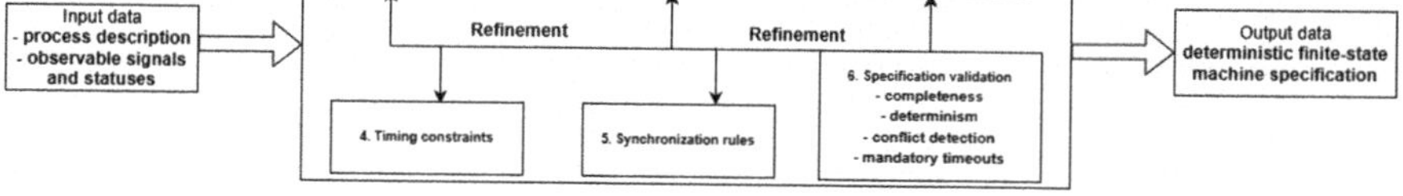

Fig. 1 Structural diagram of the method for the formal description of control logic

procedure. For each event, a formal registration condition is specified. This condition links the event to a set of observable indicators and defines at which signal and status values the event is considered to have occurred. The event dictionary is formed with reference to the already constructed state dictionary. For each state, it is analysed which events can be registered in this state and which of them are admissible.

At the third step, transitions and transition conditions are defined. For each "source state—event" pair, it is determined whether a state change is possible and which target state should be reached. If a transition is allowed, a transition condition is specified. The transition condition refines the combinations of indicators and variable values for which the transition is considered active. The condition includes, in particular, additional permissive conditions and checks of resource availability. Transitions are recorded in the form of a table in which each row contains the source state, event, condition, target state and actions performed during the transition. At this stage, the basic structure of the finite-state machine is formed, describing the nominal behaviour scenarios. Transitions related to deviation handling can be specified immediately or added at the next step after analysing timing constraints and synchronization.

At the fourth step, timing constraints are introduced. For each state and for selected events, maximum admissible waiting times are defined. These times determine how long the system may remain in a state while waiting for a confirming event or resource readiness. For each constraint, a reaction to exceeding the limit is specified. As a rule, this is the generation of a separate event that initiates a transition to a deviation-handling state, a transition to an operator-waiting mode or a stop with safe completion of the current stage. Timing constraints are formalized as a separate set of rules. These rules refine the transitions that have already been defined and extend the event dictionary with "waiting timeout" events.

At the fifth step, synchronization rules are defined. At this stage, situations are identified in which an operation depends on the simultaneous fulfilment of several conditions or on a coordinated state of several nodes of the production cell. Synchronization events are formed for such situations. They describe the fact of joint fulfilment of the required conditions, for example simultaneous readiness of the stacking and gripping zones, presence of both a bag and a pallet, or release of several technological zones. Synchronization rules can be specified as additional transition conditions or as separate intermediate states in which the system waits for all required conditions to be met. As a result, the structure of the finite-state machine is refined for transitions between states related to parallel equipment actions.

At the sixth step, specification validation is performed. At this stage, the constructed deterministic finite-state machine is analysed with regard to the state and event dictionaries, transitions, timing constraints and synchronization rules. Validation includes several groups of checks. The completeness of the description is checked, that is, the presence of outgoing transitions for significant combinations of states and events and the reachability of key working-cycle states. The determinism of transition selection is checked. For each state, it is analysed whether there exist situations in which the conditions of several transitions may be true simultaneously. If conflicts are detected, the corresponding rules are identified and adjusted. In addition, the presence of mandatory waiting timeouts is checked for states in which delayed confirmation or resource waiting is allowed. If necessary, states and transitions for which waiting and deviation-handling rules are not fully specified are marked.

The result of applying the method is a specification of logical control in the form of a deterministic finite-state machine for which the state and event dictionaries, transition table, timing constraints and synchronization rules are defined, and an initial check of completeness and determinism is performed. This specification can be used as a basis for developing control programs and for building models of the discrete behaviour of a palletizing cell for packaged agricultural products.

3 Results

3.1 Description of the Palletizing Cell and Technological Process

The proposed method is applied to an automated palletizing cell for packaged agricultural products. The cell is located at the end of the packaging line and is responsible for placing bags with feed, seeds or mineral fertilizers on pallets for subsequent transport and storage. Figure 2 shows a simulation model of the palletizing cell. The model illustrates conveyor lines supplying bags and pallets, as well as two robotized stations that perform gripping, transfer and stacking operations.

Bags with product arrive on the indeed conveyor from the preceding packaging operations. The presence of a bag in the gripping zone is detected by a sensor. Pallets are supplied to the pallet station, where the presence of a pallet and the correctness of its position are monitored. Depending on the line configuration, the station may support several pallet positions or several stacking levels. The robotized stations perform bag gripping, transfer and stacking. Their readiness is specified by a set of statuses such as "systems homed", "automatic mode enabled", "no active faults", "gripper ready to operate". Safety devices, for example light curtains or door limit switches, generate interlock signals that inhibit motion when the protected zone is violated.

The technological cycle of the cell is represented as a sequence of discrete steps. In the nominal mode, the cycle includes waiting for a pallet in the stacking zone, waiting for a bag in the gripping zone, gripping the bag, transfer to the pallet, stacking according to the specified pattern, confirmation of stacking result and transition to the next bag. Each step relies on confirming events that indicate successful completion of the action. Examples include events "bag gripped", "pallet position reached", "stacking completed". For each step, timing constraints are specified that define the maximum admissible waiting time for resource readiness or motion completion.

In addition to the nominal sequence, the cell must correctly handle deviations. Typical deviations include absence of a pallet or bag, occupation of the stacking zone, loss of the bag during transfer, robot fault messages and activation of safety interlocks. In such situations, the control system transfers the cell to alternative states

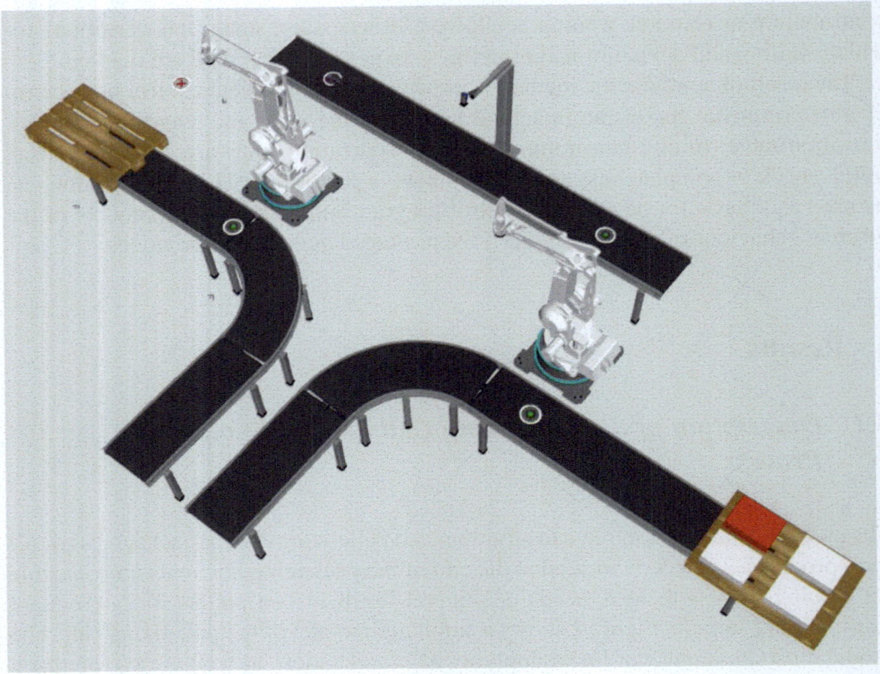

Fig. 2 Simulation model of the palletizing cell for packaged agricultural products

such as "resource waiting", "operator intervention required", "emergency stop with safe positioning". The required behaviour under deviations is determined by a combination of technological requirements and safety rules adopted at the agricultural or processing plant.

In this study, the cell is considered at the level of logical control. Robot kinematics, trajectory planning and closed-loop control of speeds and positions belong to local control tasks. These tasks are assumed to be implemented in the controller programs and are not included in the method. The focus is on the discrete structure of the cycle: the conditions for starting each step, the events that confirm its completion and the rules for transitions between states are described. This level of abstraction is sufficient to construct a deterministic specification of the palletizing process in the form of a finite-state machine and to demonstrate the steps of the proposed method.

3.2 *Formation of the Stage Dictionary*

At the first stage of applying the method to the palletizing cell, a state dictionary was created. The source information for it was the description of the technological cycle, the equipment structure and the list of typical deviations characteristic of robotized palletizing of packaged agricultural products. The task was to represent the behaviour of the cell as a finite and closed set of logical states, each having an unambiguous technological meaning and a defined role in the specification.

When forming the state dictionary, the palletizing cycle was divided into several groups of states. The first group includes the initial state and the states corresponding to the main stages of the working cycle. Examples are the initial readiness state of the cell, the "load gripping" state and the "load transfer to pallet" state. These states describe the nominal behaviour of the cell when all resources are available and no deviations occur. Invariants are specified for these states that reflect the current configuration of equipment and resources. For example, for the initial readiness state the invariants include the absence of active faults, the equipment being in its home position and the absence of a gripped load.

The second group consists of states for waiting for resources and permissive conditions. They are used for an explicit description of situations where execution of the next step of the cycle is impossible because a pallet, a unit load or the stacking zone is not available. Examples are the states "waiting for pallet" and "waiting for part (bag) in the gripping zone". The invariants of these states record that the sensor for the corresponding resource is not active, manipulator motion is blocked and the system is in a safe waiting position. Introducing waiting states makes it possible to specify formal rules for the response to extended waiting, including transitions to timeout states and to operator-intervention modes.

In the full state dictionary, states for deviation and fault handling are distinguished separately. They include, for example, "gripping error handling", "stacking error handling", "operator waiting after fault" and "emergency stop with safe positioning". These states are not shown in the fragment in Table 1, but they are taken into account when forming the transition table and validating the specification.

Table 1 Fragment of the state dictionary for the palletizing cell

Identifier	State name	Role	Purpose	Invariants
S0	Cell initial state	Working step	Cell ready for palletizing cycle	No faults; home position; no bag gripped; interlocks closed
S1	Waiting for pallet	Resource waiting	Waiting for a pallet in placement zone	No pallet signal; motion blocked; safe waiting pose
S2	Waiting for bag in pick-up zone	Resource waiting	Waiting for bag in pick-up zone	Bag not gripped; no motion command; manipulator in pick-up pose
S3	Bag gripping	Working step	Bag gripping in progress	Gripping command active; motion limited to pick-up zone
S4	Bag transfer to pallet	Working step	Transfer of the gripped bag to pallet zone	Bag gripped; pallet present; path to pallet zone free

For each state in the dictionary, an identifier, a name and a brief textual description of its purpose are specified. In addition, invariants and, where necessary, actions on state entry and exit are recorded. Such actions include resetting waiting time counters, switching individual mechanisms on or off and registering messages in the event log. For analysis convenience, each state is assigned a role in the specification structure. The roles used are working step, resource waiting, deviation handling and fault state. This makes it possible to check the completeness of the cycle, including the presence of initial and final states and the existence of paths from deviation and fault-handling states to safe modes.

A fragment of the resulting state dictionary is presented in Table 1. It includes the initial state of the cell, key working-cycle states and resource-waiting states. The tabular representation provides a compact record of the characteristics of each state and serves as a basis for subsequent definition of admissible events and transition rules between states.

3.3 Formation of the Event Dictionary and Registration Conditions

At the next stage, an event dictionary was created for the palletizing cell. The source data were the list of sensor signals and equipment statuses, as well as the description of situations that are significant for choosing the further behaviour of the cell. The aim was to identify a finite set of events used by the control logic and to define a formal registration condition for each of them. This made it possible to link technological facts represented in the simulation model of the cell (Fig. 2) with elements of the deterministic finite-state machine specification.

The events were divided into several groups. The first group includes events that confirm the completion of technological actions. They are used to advance along the nominal palletizing cycle. Typical examples are "bag gripped", "pallet position reached" and "placement completed". For each event in this group, it is specified which action is considered completed and which indicators are sufficient for confirmation. For example, the "bag gripped" event is registered when two conditions are satisfied simultaneously: the gripper sensor signal is active and the manipulator is located in the gripping zone while no motion command is executed.

The second group comprises events that characterize the availability of resources and permissive conditions. These events determine whether the next step of the cycle can be executed. Examples are "pallet in stacking zone", "bag in gripping zone", "stacking zone free" and "safety zone closed". For each such event, its source in the automation system is defined, for example a particular sensor on the conveyor or on the pallet station, and a registration condition is specified. Thus, the "bag in gripping zone" event is registered when the bag presence sensor is active and the "bag gripped" signal is absent. The "pallet in stacking zone" event is registered when the pallet presence signal is active and the pallet is in the correct position according to the locating sensor.

Deviation and fault events are distinguished as a separate group. This group includes, for example, "gripping error", "placement error", "safety interlock activated" and "robot fault". Such events are initiated by robot diagnostic signals, by the state of safety interlocks and by logical conditions that reflect the impossibility of executing a command. Registration conditions for this group describe which combinations of diagnostic indicators are critical and require a transition to deviation-handling or fault states. For instance, the "safety interlock activated" event is registered when the safety circuit is open and a motion command is active.

The fourth group consists of timeout events. They describe the expiry of the admissible waiting time for confirmation or for resource appearance. For each waiting state, a maximum interval is specified, after which the corresponding timeout event is registered. Examples are "pallet waiting timeout", "bag waiting timeout" and "motion completion timeout". These events are closely related to the timing constraints defined at the stage of forming waiting rules and are used to trigger transitions to deviation-handling states or to operator-intervention modes.

For each event in the dictionary, an identifier, a name, a short description, a source in the automation system and a registration condition are specified. The registration condition is formulated in terms of observable signals and statuses. It defines which indicators must be true for the event to be considered occurred and which indicators must be false. In some cases, observable equivalents are used when the required fact cannot be registered by a single physical signal. In such situations, the event is formed from a combination of available indicators. For example, the "stacking zone free" event can be defined as the joint fulfilment of the following conditions: no bag according to the presence sensor, no active placement commands and no recorded result of the previous operation.

3.4 Transition Table, Timing Constraints and Synchronization Rules

At the next stage, a transition table of the finite-state machine was constructed for the palletizing cell. The input data were the state dictionary (Table 1), the event dictionary (Table 2), and the previously defined timing constraints and synchronization rules. The task was to determine, for each "source state—event" pair, whether a transition is allowed, what the target state is, under which condition the transition is activated, and which actions are executed during the transition. The resulting table specifies the discrete structure of the bag palletizing cycle and serves as a basis for subsequent implementation of the control logic in control programs.

The transition table was formed step by step. First, transitions for the nominal operating scenario of the cell were described. For this purpose, states related to the working cycle and confirmation events were used. For example, when event EV_

Table 2 Fragment of the event dictionary for palletizing cell

Identifier	Event name	Event type	Short description	Registration condition
EV_GRIP_OK	Bag gripped	Action confirmation	Successful bag gripping	Gripper sensor active; manipulator in pick-up zone; no motion command
EV_PLACE_OK	Placement completed	Action confirmation	Bag placement completed	Placement complete signal active; no manipulator movement
EV_TO_PLT	Pallet waiting timeout	Timeout	Pallet waiting time exceeded	Pallet waiting timer elapsed; no pallet presence signal

START_CYC occurs in state S0 (cell initial state), the system moves to state S1 (waiting for pallet). In state S1, when event EV_PLT_READY (pallet ready for placement) is registered, the system moves to state S2 (waiting for bag in pick-up zone). The transition condition additionally requires that the placement zone is free, the safety zone is closed, and no active faults are present. Similarly, in state S2, when event EV_WP_READY (bag in pick-up zone) occurs, the system moves to state S3 (bag gripping). For this transition the manipulator must be in the gripping standby pose and must not perform other motions.

Then transitions related to completion of gripping, transfer, and placement operations were added. In state S3, when event EV_GRIP_OK (bag gripped) occurs, the system moves to state S4 (bag transfer to pallet). The transition condition requires that the gripper sensor signal is active and the pick-up zone is free. In state S4, after motion completion and event EV_PLT_POS (pallet position reached), the system moves to state S_PLACE, which describes bag placement on the pallet. After placement completion and registration of event EV_PLACE_OK, the system moves either to a result-confirmation state or back to a state waiting for the next bag, depending on the adopted cycle structure. For each transition, the actions executed on its activation are specified. Typical actions include starting manipulator motion, resetting waiting counters, updating the number of bags placed on the pallet, and writing service records to the event log.

A separate set of transitions is defined for resource-waiting states and timeout events. In state S1 (waiting for pallet) a transition is specified for event EV_TO_PLT (pallet waiting timeout). This transition leads to state S_PLT_TIMEOUT, which belongs to deviation handling. The transition condition fixes that the specified waiting interval has expired and event EV_PLT_READY has not been registered. Subsequent transitions from state S_PLT_TIMEOUT move the cell either to an operator-intervention mode or to safe cycle termination. Similar transitions can be defined for states waiting for a bag and waiting for motion completion. As a result, timing constraints become part of the finite-state machine model. They determine not only admissible waiting intervals but also the further behaviour of the system when these intervals are violated.

Transitions related to deviation and fault handling are formed on the basis of deviation events. For example, when event EV_GRIP_ERR (gripping error) occurs in state S3, the system moves to state S_GRIP_ERR, which describes gripping error handling. In this state, motion commands are disabled and actions for safe completion of the operation and indication of the deviation cause are executed. When safety interlocks are activated, a transition from any working state to an emergency state with safe positioning is performed. For such transitions, invariants of the target states include motion disabling and preservation of information about the fault cause. Additional transitions are defined from deviation-handling states to operator-waiting states and then back to normal operating modes after the cause has been removed.

Synchronization rules are reflected in the transition table through conditions in which several indicators or several events are checked simultaneously. For example, a transition from the initial readiness state to the state waiting for a bag may be allowed only when the following conditions are met at the same time: a pallet is present in the placement zone, the manipulator travel path is free, and the safety zone is closed. Such requirements are recorded in the transition condition as the joint fulfilment of several logical predicates. In more complex cases, synchronization is represented by separate intermediate states in which the system waits for a coordinated state of several units, for example simultaneous readiness of two robotized cells or several placement zones.

Formally, the transition table is represented as a set of rows, each containing a source state, an event, a condition, a target state, and actions on transition. A fragment of this table is presented in Table 3. The tabular representation facilitates analysis of specification correctness. It is used to check completeness, that is, the presence of the required set of transitions for all significant combinations of states and events, and to check determinism. For each state it is analysed whether there exist situations in which several rows of the table can be activated simultaneously. If such conflicts are detected, transition rules are corrected, for example by clarifying conditions or introducing priorities.

The graphical representation of the finite-state machine built from the transition table is shown in Fig. 3 and is used for visual analysis of the control-logic structure.

3.5 Validation Results of the Specification

The final stage of applying the method to the bag palletizing cell was validation of the obtained control-logic specification. The validation was carried out using the state dictionary (Table 1), the event dictionary (Table 2), the transition table (Table 3), and the graphical representation of the deterministic finite-state machine (Fig. 3). The main goal was to check the completeness of the description, the determinism of transition selection, and the consistency of the automaton structure with the technological cycle of the filling and palletizing cell for agricultural products.

Table 3 Fragment of the transition table for the palletizing cell

Source state	Event	Condition	Target state	Actions on transition
S0	EV_ START_ CYC	Cell is in the initial state; no active faults	S1	Initialize counters and timers; enable cell readiness
S1	EV_PLT_ READY	Pallet present in placement zone; zone free; safety zone closed	S2	Enable manipulator motion; move manipulator to the pick-up pose
S1	EV_TO_ PLT	Pallet waiting timer elapsed; event EV_PLT_ READY not registered	S_PLT_ TIMEOUT	Block motion; set "no pallet" indication; log timeout event
S2	EV_WP_ READY	Bag in pick-up zone; "bag gripped" signal inactive	S3	Start gripping cycle; reset bag waiting timer

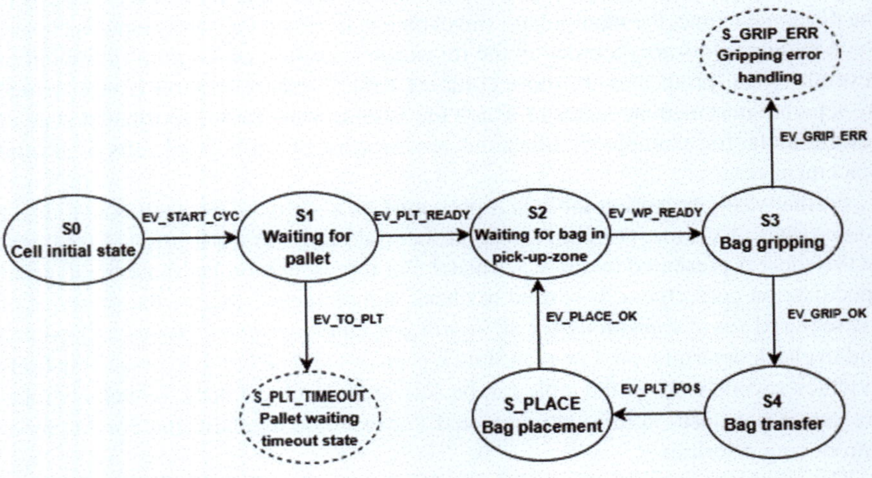

Fig. 3 Fragment of the finite-state machine for the palletizing cell

The first step was to check the completeness of the specification. The analysis was performed by groups of states. For states belonging to the working cycle, the presence of incoming and outgoing transitions for key events was verified, such as bag arrival in the pick-up zone, gripping completion, and placement completion. For resource-waiting states it was analysed whether transitions were specified both for the occurrence of the confirming event (pallet or bag appearance) and for the expiry of the waiting timeout. For deviation-handling and fault states the existence of routes back to safe modes and the possibility of resuming the cycle after the cause had been eliminated were checked. During the analysis several incomplete scenarios were identified in which, for certain combinations of events, the system could remain in a waiting state without an explicit transition. In particular, it was necessary to add transitions from the "waiting for bag" state for the case when the bag does not appear but the pallet is replaced by the operator, and to introduce an explicit transition from the gripping-error handling state to the operator-waiting state.

At the next step the determinism of transition selection was analysed. For each state, all rows of the transition table originating from this state were considered. It was checked whether there exist combinations of observable indicators for which the conditions of several transitions can become true simultaneously. For working-cycle states such an overlap was found, for example, between the transition triggered by the "bag in pick-up zone" event and the transition triggered by the gripping error event. In the initial version, both situations could be interpreted for the same set of signals. To remove this ambiguity, the registration conditions for the events and the transition conditions were refined. For the transition triggered by the gripping error event, an additional check of a diagnostic indicator was included in the condition, and for the transition triggered by the "bag in pick-up zone" event a requirement of absence of an active error code was added. Similar corrections were made for transitions related to placement completion and activation of safety interlocks.

Timing constraints and the associated timeout events were checked separately. For each waiting state it was analysed whether a maximum waiting interval was specified and whether there was a corresponding transition for the timeout event. In several cases the initial description assumed that long waiting would be interrupted by the operator without a formally defined timeout. Within the proposed method such situations were replaced by explicit rules. For the pallet-waiting and bag-waiting states the EV_TO_PLT and EV_TO_WP events were introduced, and corresponding transitions to deviation-handling states were added to the transition table. This made it possible to formally define the system response to prolonged absence of resources and to ensure uniform behaviour of the cell in deviation situations.

The graphical representation of the finite-state machine (Fig. 3) was used for visual analysis of the control-logic structure. The graph was used to evaluate the reachability of key working-cycle states and the existence of paths from deviation-handling and fault states. As a result, transitions from the emergency stop with safe positioning state were refined. In the initial version, the return to normal mode was assumed through a general fault-reset scenario. The specification was extended by explicit transitions that connect this state with the operator-waiting state and then with the initial state of the cell after the required checks have been completed.

The validation showed that the formal specification of the control logic for the bag palletizing cell makes it possible to identify and eliminate a number of potential design errors already at the level of the logical model, before implementation in programmable logic controllers. During the analysis, incomplete resource-waiting scenarios, possible conflicts of transition conditions, and implicit assumptions about system behaviour under deviations were detected. Their elimination led to a more transparent automaton structure and to explicitly defined rules for all significant combinations of states and events. This confirms the feasibility of applying the proposed method to describe the control logic of discrete palletizing cells in agro-industrial production, since it provides a way to systematically verify the correctness of design decisions before commissioning.

4 Discussion

The results of applying the method to the bag palletizing cell in agro-industrial production show that a formal description of control logic in the form of a deterministic finite-state machine makes it possible to reorganize work with discrete behaviour of technological systems. The specification, which includes a state dictionary, an event dictionary, a transition table, timing constraints, and synchronization rules, defines a consistent description of the palletizing cycle. This description links technological requirements, equipment structure, and deviation response rules in a single formal representation suitable both for analysis and for subsequent implementation in control programs.

One of the key effects is a reduction in fragmentation of control logic. In typical projects of filling and palletizing cells, the behaviour of the system is formed separately: in controller programs written in different languages, in configuration of conveyor and robotic equipment, in operator panel settings, and in fault-handling rules. As a result, engineers rely on heterogeneous representations, and change coordination is carried out mainly at the commissioning stage. In the proposed approach, discrete behaviour is first defined explicitly through a finite set of states and events. The transition table records which combinations of events and conditions lead to a state change and which keep the system in the current mode. This makes it possible to detect gaps and contradictions before developing programs in the controller environment and before putting the equipment into operation.

Another advantage is the possibility of systematic correctness checking. Validation of the specification, described in the previous section, showed that even on a relatively simple bag palletizing cell non-obvious scenarios are revealed. They are related to resource waiting, combinations of deviation events, and long delays. In real conditions such situations often appear only at the commissioning stage or during operation under atypical modes of the line. The formal representation of logic makes it possible to check in advance the completeness of transition sets, the presence of timeouts for critical waiting periods, and the absence of conflicting conditions. This reduces the risk that the cell will remain in an intermediate state or respond to a deviation in an ambiguous way.

The proposed approach is compatible with existing tools and standards for automation programming. State diagrams, sequential schemes, graphs, and SFC languages are already used to describe discrete behaviour in equipment for storage, transport, and processing of agricultural products. However, in practice these models often remain at the documentation level and do not form a single specification that serves as a common reference point for all project participants. In the method considered here, the finite-state machine is not only a visual aid. It defines the structure of states and transitions, which can later be used to synthesize programs in LAD and ST languages or to design SFC steps. Thus, the formal model does not replace existing controller programming languages but complements them. It introduces an intermediate level of description at which the control logic is checked and agreed upon before implementation.

An important aspect is the connection with modelling and digital twins of agro-industrial production cells. In this work, a simulation model of the palletizing cell was used as input data. The formal specification of the finite-state machine can be used as a data source for such models. The states of the automaton correspond to operating modes of the cell, and the events correspond to observable facts in the simulation environment. This makes it possible to align the behaviour of the virtual and real cell and to use unified scenarios for algorithm testing and commissioning trials. In addition, the deterministic transition structure simplifies the generation of test scenarios that cover not only the nominal cycle but also deviation situations, waiting timeouts, and activation of safety interlocks.

At the same time, the limitations of the method should be taken into account. Formalization of control logic requires additional effort at the design stage. It is necessary to explicitly define the set of states and events, refine registration conditions, construct the transition table, and perform validation. For relatively small cells such as the bag palletizing example considered, these costs are justified by a reduction in the number of commissioning iterations and simplification of subsequent modernization. When moving to large lines with several interconnected cells and complex routing of product flows, the volume of the specification increases significantly. In this case, tool support becomes necessary. Editing and visualization tools for finite-state machines, automatic checking of determinism and completeness, and mechanisms for integrating the specification with controller code and simulation systems are required.

Another limitation is that in this paper the method is demonstrated on a bag palletizing cell and is primarily aimed at discrete transport and placement operations of product containers. Although the finite-state formalism used is applicable to other discrete tasks in the agro-industrial sector and agricultural machinery, more complex processes will require adaptation. This concerns, for example, coordination of several robotic cells, dynamic redistribution of tasks between them, and combination of discrete logic with control of technological regimes such as drying, dosing, or thermal treatment. In such cases, a hierarchical organization of states or a combination of several automata describing different levels of the system may be required.

A promising direction for further development is deeper integration of the formal control-logic specification with information models of production processes in the agro-industrial complex. The structure of the finite-state machine, represented as dictionaries and tables, can be mapped to a data model and used as a single source of information for various engineering tasks. This opens up possibilities for closer linkage of control logic with planning and dispatching systems, as well as with production data analysis tools. In addition, the specification constructed using the proposed method can serve as a basis for semi-automatic generation of control programs, for building verifiable models in simulation environments, and for assessing the effectiveness of control decisions before implementation on real equipment.

In general, the experience of applying the method to a robotic bag palletizing cell in agro-industrial production shows that a deterministic finite-state machine with explicitly defined states, events, transitions, timing constraints, and synchronization rules can be regarded as a convenient form of control-logic specification. Such a specification increases the transparency of design decisions, simplifies interaction between automation engineers, process engineers, and simulation model developers, and creates prerequisites for further integration of control logic with digital twins and the information infrastructure of an agro-industrial enterprise.

5 Conclusion

This paper proposes an approach to the formal description of control logic for discrete technological processes in agro-industrial production, using a bag palletizing cell as an example. The control logic is defined as a deterministic finite-state machine that includes a state dictionary, an event dictionary, a transition table, timing constraints, and synchronization rules. Application of the method to a robotic palletizing cell has shown that such a specification makes it possible to link technological requirements, equipment structure, and deviation response rules within a single formal representation suitable for analysis, modelling, and subsequent implementation in control programs.

The performed validation of the specification demonstrated the ability to detect incomplete resource-waiting scenarios, potential conflicts between transition conditions, and implicit assumptions about system behaviour already at the design stage. This creates prerequisites for reducing the number of commissioning iterations, increasing the transparency of design decisions, and simplifying the modernization of discrete filling and palletizing cells. At the same time, formalization of control logic requires additional effort and tool support, especially when describing large production lines with several interconnected cells and complex routing of product flows.

A promising direction for further research is the integration of the formal finite-state machine specification with information models of production processes and simulation systems in the agro-industrial sector. Mapping the automaton structure to a data model and using it as a single source of information for various engineering tasks will strengthen the link between control logic, digital twins, and the information technology infrastructure of the enterprise. It will also provide a basis for developing tools for generation and verification of control programs for discrete sections of storage, transport, and processing of agricultural products.

References

1. Konfo, T.R.C., Djouhou, F.M.C., Hounhouigan, M.H., Dahouenon-Ahoussi, E., Avlessi, F., Sohounhloue, C.K.D.: Recent advances in the use of digital technologies in agri-food processing: a short review. Appl. Food Res. **100329** (2023). https://doi.org/10.1016/j.afres.2023.100329
2. Sridhar, A., Ponnuchamy, M., Kumar, P.S., Kapoor, A., Nguyen Vo, D.V., Rangasamy, G.: Digitalization of the agro-food sector for achieving sustainable development goals: a review. Sustain. Food Technol. **783–802** (2023). https://doi.org/10.1039/d3fb00124e
3. Escribà-Gelonch, M., Liang, S., van Schalkwyk, P., Fisk, I., Long, N.V.D., Hessel, V.: Digital twins in agriculture: orchestration and applications. J. Agric. Food Chem. **10737–10752** (2024). https://doi.org/10.1021/acs.jafc.4c01934
4. Chauhan, A., Brouwer, B., Westra, E.: Robotics for a quality-driven post-harvest supply chain. Curr. Robot. Rep. **39–48** (2022). https://doi.org/10.1007/s43154-022-00075-8
5. Mo, F., Ugarte Querejeta, M., Hellewell, J., Rehman, H.U., Illarramendi Rezabal, M., Chaplin, J.C., Sanderson, D., Ratchev, S.: PLC orchestration automation to enhance human–machine integration in adaptive manufacturing systems. J. Manuf. Syst. **172–187** (2023). https://doi.org/10.1016/j.jmsy.2023.07.015
6. Xiong, J., Zhu, G., Huang, Y., Shi, J.: A user-friendly verification approach for IEC 61131-3 PLC programs. Electronics. **572** (2020). https://doi.org/10.3390/electronics9040572
7. Babakov, R.M., Barkalov, A.A.: Determining the area of efficient application of a microprogrammed finite-state machine with Datapath of transitions. Cybern. Syst. Anal. **366–375** (2018). https://doi.org/10.1007/s10559-018-0038-8
8. Kuznecov, N.A., Antonov, S.V.: Automata-based programming practice for networked control systems development. Vestnik kompyuternykh i informacionnykh tekhnologij. **37–43** (2024). https://doi.org/10.14489/vkit.2024.03.pp.037-043
9. Zhang, R., Cai, K., Gan, Y., Wang, Z., Wonham, W.M.: Supervision localization of timed discrete-event systems. Automatica. **2786–2794** (2013). https://doi.org/10.1016/j.automatica.2013.05.015
10. Kaczmarek, W., Borys, S., Panasluk, J., Siwek, M., Prusaczyk, P.: Experimental study of the vibrations of a roller shutter gripper. Appl. Sci. **9996** (2022). https://doi.org/10.3390/app12199996
11. Kholopov, V.A., Klyagin, M.M., Ogorelcev, R.M., Mishina, V.A.: Method for designing a control system for the box palletizing process based on finite state machine and relational data model. Sovremennye naukoemkie tekhnologii. **62–71** (2025). https://doi.org/10.17513/snt.40423

The manufacturer's authorised representative in the EU is Springer Nature Customer Service Centre GmbH, Europaplatz 3, 69115 Heidelberg, Germany. If you have any concerns regarding our products, please contact ProductSafety@springernature.com

Printed and bound by CPI Group (UK) Ltd, Croydon, CR0 4YY
21/09/2026
02219993-0006